Innovating for The Circular Economy

Driving Sustainable Transformation

I0027316

Edited by
Rachna Arora, Dieter Mutz, and
Pavithra Mohanraj

CRC CRC Press
Taylor & Francis Group
Boca Raton London New York

CRC Press is an imprint of the
Taylor & Francis Group, an **informa** business

Designed cover image: Illustrator - Agalya (Founder - Noun & Verb)

First edition published 2023
by CRC Press
6000 Broken Sound Parkway NW, Suite 300, Boca Raton, FL 33487-2742

and by CRC Press
4 Park Square, Milton Park, Abingdon, Oxon, OX14 4RN

CRC Press is an imprint of Taylor & Francis Group, LLC

© 2023 selection and editorial matter, Rachna Arora, Dieter Mutz, and Pavithra Mohanraj; individual chapters, the contributors

ISBN: 9781032063348 (hbk)
ISBN: 9781032063386 (pbk)
ISBN: 9781003201816 (ebk)

DOI: 10.1201/9781003201816

Typeset in Times
by Deanta Global Publishing Services, Chennai, India

Innovating for The Circular Economy

Systemic change is required to move to a circular economy (CE) model which can meet the demands of a growing population in a manner that is decoupled from resource use and waste generation. This book takes a deep dive into the innovation aspect of the circular economy (CE), with a specific focus on India as a geography, where the transformation to a circular economy is underway.

How a developing country like India is tackling the complexities of the transformation and creating innovative solutions is showcased in this book through many practical examples and inspirational case studies. The book lays out the foundations for mainstreaming resource efficiency (RE)/CE in India, and covers innovation led by businesses and start-ups, along with the innovative policies, financing, and collaborative models required to spur and accelerate circular economy approaches, while also providing linkages to the international context.

Features:

- Provides insight into the role of innovation in the circular economy transition;
- Helps to develop and facilitate adoption of resource-efficiency policy and strategy with particular focus on key resource sectors and waste streams;
- Treats the circular economy as a holistic approach across the entire lifecycle, and places emphasis on upstream interventions and systems change;
- Examines the current context of COVID-19 and its impact on circular economy models and practices;
- Touches upon how the EU-based approach was adapted and contextualised significantly to work in the unique Indian landscape.

This book is aimed at students, researchers, and professionals engaged in the domains of circular economy, sustainability, business innovation, environmental studies, natural resources management, and environmental and resource conservation policy.

Contents

About the Editors

Dr Rachna Arora is working as a team leader in the EU-Resource Efficiency project of the Deutsche Gesellschaft für Internationale Zusammenarbeit (GIZ) GmbH, Bonn, Germany, and as a team leader on the GIZ Circular Economy solutions preventing Marine litter project. She has been working with GIZ for the last 14 years with relevant ministries and government departments towards environmental policy formulation and its implementation. She is a TEDx speaker on the circular economy.

She holds a doctorate degree in environmental chemistry from the Indian Institute of Technology (IIT) Roorkee, Uttarakhand, India. She was a member of the interdepartmental committee set up by NITI Aayog (policy Think Tank, GoI) on Resource Efficiency strategy, NITI Aayog's committee set up by the Ministry of Information Technology and Communication (MeitY), the GoI Circular Economy Action plan on e-waste, MNRE's Committee on Solar Panel Recycling, and the Ministry of Steel committee on scrap recycling. She is also a working group member of the FICCI Circular Economy Committee, supporting industry research and dialogues on the circular economy.

Dieter Mutz's professional background is in sanitary engineering, and he has more than 30 years of work experience with bi- and multilateral development organisations (GTZ, KfW, SECO, DEZA, UNIDO, the World Bank, IDB) with assignments in Latin America, the Middle East, and South Asian regions. His professional expertise lies in eco-efficient resource management and integrated solid waste and wastewater management.

His international experiences with resource efficiency and circular economy derive from his assignment as team leader for the EU-financed project called 'Resource Efficiency Initiative for India' (January 2017–December 2020) and as Director of the Indo-German Environment Partnership Programme (IGEP), jointly implemented by the Indian Ministry of Environment, Forests & Climate Change (MoEFCC) and GIZ (Deutsche Gesellschaft für International Zusammenarbeit), financed by the German Ministry for Economic Cooperation and Development (August 2010–September 2015). This programme had a focus on sustainable urban and industrial development, land and resource management, as well as environmental protection and climate change mitigation.

In addition, since January 2016, Dieter holds a position as lecturer at the University of Applied Sciences, Northwestern Switzerland (FHNW), an academic and research organisation he had already been affiliated with from 1999 until 2010. His academic focus lies on integrated waste and resource management in Europe and countries in transition.

Dieter holds a doctorate from the University of Karlsruhe where he studied Sanitary Engineering. He has had long-term assignments in Thailand, Nepal, and Ecuador as well as India.

Pavithra Mohanraj is the co-founder of Infinitive, a young advisory firm working to accelerate the adoption of the circular economy in India through communications, capacity building, business advisory, and ecosystem development projects. Infinitive has played a key role in shifting mindsets and shaping the evolving conversation around the circular economy in India.

Pavithra graduated from the National University of Singapore with a Bachelor of Science (with Merit). She has over a decade of experience leading projects in sustainability and Circular Economy across Singapore and India, and has worked with industry, as well as bilateral and multilateral agencies. Most recently, Pavithra has also completed a year-long stint at the United Nations Environment Programme (UNEP) as a Consultant supporting projects under the Sustainable and Circular Textiles portfolio of the UNEP India Office.

Contributors

Manisha Anantharaman
Chatham House/Saint Mary's College of California, USA

Rachna Arora
Deutsche Gesellschaft für Internationale Zusammenarbeit (GIZ) GmbH – India, India

Abhijit Banerjee
O. P. Jindal Global University, India

Ashish Chaturvedi
Deutsche Gesellschaft für Internationale Zusammenarbeit (GIZ) GmbH – India

Anirban Ghosh
Mahindra Group, India

Narendra Prasad Kolary
Infinite Perspective Research and Advisory Private Limited, India

Pavithra Mohanraj
Infinite Perspective Research and Advisory Private Limited, India

Dieter Mutz
University of Applied Sciences and Arts, Northwestern Switzerland (FHNW)

Katharina Paterok
Deutsche Gesellschaft für Internationale Zusammenarbeit (GIZ) GmbH, Germany

Reva Prakash
Deutsche Gesellschaft für Internationale Zusammenarbeit (GIZ) GmbH – India, India

Srikanth Prakash
Deakin University, Waurn Ponds Campus, Australia

Swati Singh Sambyal
Independent Waste and Circular Economy Expert, India

Shreya Sankar
Mahindra Group, India

Patrick Schroeder
Chatham House, UK

Stefan Šipka
European Policy Centre, Belgium

Mayuri Wijayasundara
Deakin University, Waurn Ponds Campus, Australia

Preface

During the Paris 2015 United Nations Climate Change Conference (COP21), countries agreed to limit global temperature rise to well below 2°C (compared to pre-industrial levels) by committing to emissions reductions. To accomplish this, the adoption of renewable energy and energy-efficiency measures is not nearly sufficient. Emissions linked with industrial production, resource consumption, and waste management also need to be drastically reduced. There is growing recognition that the overall economic model needs to be shifted from the 'take-make-dispose' linear economy model to a resource-efficient, restorative, and regenerative one, by imbibing circular economy (CE) principles. This was further reiterated in the Glasgow COP26 at the end of 2021 with the prioritisation of areas such as innovation, sustainable consumption, product life extension, and accelerating transformation to a circular economy.

A circular economy would require a systemic transition involving replacement of the end-of-life concept with the reduction and elimination of wastes, use of renewable energy and inputs, eco-design, product life extension models, and maximising the efficiency of use of all input materials. Enabling the adoption of these approaches across all lifecycle stages will require innovation in policy and governance, innovative financial and market instruments, along with support to start-ups, collaboration, digitalisation, and social innovation for just transition. The smart design of materials, products, and systems would require a collaborative ecosystem for addressing resilience in value chains through the active engagement of all stakeholders including government, research institutions, and the private sector.

Currently, CE innovation is quite advanced in Europe with rapidly evolving research and availability of in-depth literature. However, in India, the CE and resource efficiency (RE) policy framework is evolving with different forms of innovation being experimented with, but this is not yet well documented. There is a lack of publications which provide a holistic view on the CE innovation landscape, challenges faced by start-ups, and the choice of financial instruments and their potential for the Indian economy. This book will address this gap by deep diving into the progress being made in India, along with reflections on the policy progress in the EU.

Innovating for The Circular Economy – Driving Sustainable Transformation provides an overview of how the RE/CE concepts have begun to be mainstreamed in India, the latest developments in CE innovation, and insights on this progress. The book focuses on innovation within businesses towards CE, including the culture and collaboration required to create the ecosystem towards transition. It also covers policy and financial instruments which play a key role in enabling the transformation, as well as the need for rethinking these in the context of the circular economy. It also focuses on the collaborative approaches required to spur, finance, and mainstream CE innovation, along with a showcase of successful start-ups in the CE space in India, and highlights the potential of socially inclusive CE innovation. [The contents of this book are largely based on research restricted to the period until December 2021, and may not reflect developments beyond this period.]

This book is intended for sustainability practitioners, students, policymakers, entrepreneurs, and researchers seeking to gain knowledge and in-depth understanding of the 'Indian way' of transitioning towards the circular economy. Through sharing of perspectives and reflections from domain experts (national and international) on the CE transition in India against the backdrop of certain global developments, this book offers a unique proposition for a variety of audiences (at the Indian and international level). It also provides an understanding of the challenges and complexities of the transition, and how some of these are being tackled in the country through policy, financial and social innovation, the involvement of the private sector and start-ups, collaborative approaches, and digitalisation.

The book showcases many practical examples and inspirational case studies, with the aim of creating an understanding of the existing models and policies, facilitating knowledge exchange on CE in India, and accelerating the pace of this innovation by developing a common vision. Every chapter in the book relates to the Indian audience, gaps and barriers, sectoral and policy issues, and solutions relevant for India, with an analysis of the international context.

The book is organised into eight chapters.

Chapter 1 begins with an analysis of the transformative potential of circular economy and applies Social, Technological, Economic, Environment, and Political (STEEP) analysis to identify the stylised set of drivers for CE transition. It provides an account of how these drivers have played out in the Indian and global contexts. The chapter also examines the progress made in recent years in mainstreaming the RE and CE agenda in India's policy framework and assesses the key policy instruments which have enabled the creation of the framework for a circular economy, with examples from the EU and India.

Chapter 2 underscores the importance of systems innovation for circular transformation and presents five types of innovation that are key for the circular economy. It presents the importance of policy coordination and multi-stakeholder collaboration to successfully navigate the complexities and challenges of transitioning to a just and sustainable economy. The chapter details the CE transition against the regenerative cultures and circular economies existing in India and South Asia and the challenges that the transition will present, given South Asian realities of rapid industrialisation, economic precarity, high rates of informal work, and enduring unemployment.

Chapter 3 discusses the crucial role of finance in supporting the transformation to a circular economy and identifies key aspects to consider with respect to financing CE initiatives. Several financing models and approaches for CE from across the world are explored, and in-depth case studies from India are analysed as well. The chapter also discusses how the concepts of economic exchange, financial consideration, and ownership are redefined in a circular economy

Chapter 4 provides an overview of the status of informal resource recovery in India and its criticality to the circular economy transition. Through selected case studies, the chapter establishes the importance of institution-driven initiatives (top-down) along with society-driven actions (bottom-up) for a transition to circular economy models at a city level, with recommendations that can be utilised for social inclusion pathways in India.

Chapter 5 begins with examining the role that private sector businesses can play in accelerating the circular economy transition and recent momentum in business activity towards CE, including in the Indian context. The second part of the chapter describes case studies from the Mahindra group that illustrate how businesses can innovate towards a circular economy model.

Chapter 6 showcases case studies of circular economy start-ups in India across different lifecycle stages and identifies key elements for the development of a circular economy innovation ecosystem in India. The chapter also illustrates the role of digitalisation as an enabler of circular economy, with a description of solutions from Europe. It further highlights the need for an enabling policy framework of action on aligning the circular economy and digitalisation agendas.

Chapter 7 focuses on emerging models of collaboration between various stakeholders in the global arena, with two in-depth innovation case studies from India, involving different sectors and at different stages of innovation. The authors examine the need for such collaborations, the advantages they offer, and good practices that enable their success.

Chapter 8 builds on the other chapters with a reflection on the uptake of circular economy models which have been affected both negatively and positively by the COVID-19 pandemic. This chapter examines the emerging trends and the impact of COVID-19 on two selected CE business models in India: plastics recycling and subscription models for mobility. Finally, the chapter highlights an example of decentralisation, which has emerged as an important strategy for consideration in rebuilding our economies in the post-COVID world.

We are quite hopeful that the insights shared in *Innovating for The Circular Economy – Driving Sustainable Transformation* will be a worthwhile read for sustainability practitioners, enabling them to design, strategise, and contribute effectively towards building a coherent CE ecosystem.

Acknowledgements

The mainstreaming of resource efficiency (RE) and circular economy (CE) has seen tremendous momentum in India since 2012. In the last 10 years, the editors and the authors of this book have witnessed the beginnings of this circular transition, anchored on innovation, at close quarters – the processes, challenges in the early stages, and emerging dialogues with various stakeholders. This book began as a conversation between the editors on the absence of consolidated documentation on this highly dynamic CE innovation landscape in India. The editors recognised the need to document reflections, challenges, and learnings on the various forms of innovation that are underpinning the CE transformation in India. This book is our attempt to capture these insights from a uniquely Indian perspective, while also identifying linkages to global developments in CE.

At the very outset, we would like to thank all the contributing authors of this book for coming on board this project with as much excitement as us. To our authors – thank you very much for being so receptive to our conceptualisation of the book contents, and for building on it with your vast expertise and insights, resulting in this well-rounded book with multiple perspectives.

We would like to express our gratitude to our publishers for seeing the value in our proposed book. Special thanks to Dr Gagandeep Singh, Senior Publisher, CRC Press, for offering valuable insights and encouragement throughout the book development process.

A FEW WORDS FROM DIETER MUTZ

I had the opportunity to work from 2017 until 2020 as GIZ team leader for the EU-financed project 'Resource Efficiency for India (EU-REI)'. During these three years, I learnt that a transition from a linear to a circular economy can't be successfully implemented by applying global recipes alone, but needs a tailor-made local agenda that reflects the historic development of the local economy, social disparity, and cultural diversity. Thank you to all my Indian friends and colleagues for helping me to understand the 'Indian way' of a transformation process better, for inspiring and accompanying me on this exciting journey over the past years – a journey for which you cannot buy a ticket anywhere. The experiences made and lessons learnt were a valuable treasure when co-editing this book.

A FEW WORDS FROM RACHNA ARORA

This book has been an opportunity to bring out the experiences that I have gathered by working in India with national and international organisations, viz. GIZ and the EU Delegation in India, partners, and experts. I would like to thank all my colleagues and friends for inspiring me to take up this assignment.

Words of appreciation will be inadequate to express my deepest admiration to my husband, Dr Raman Arora, and to my kids, Ritkriti and Riyaarth, for letting me extend my work hours beyond the office, especially during COVID-19. Finally, no words are ever adequate to express heartfelt veneration to my parents, my sister, and my loving in-laws, who stand by me firmly to keep myself engaged in higher pursuits.

A FEW WORDS FROM PAVITHRA MOHANRAJ

I would like to thank my partner, Naren Kolary, for being my sounding board and helping structure my myriad thoughts into coherent content, and for reading every word of this book with as much attention as I did, to eke out every extra bit of nuance. I would also like to thank my family – my parents, for their support and unflagging work ethic, which is my inspiration every single day, and my brother, who has been alternately providing encouragement and humorous interludes during the development of this book. Finally, a huge thanks to my mentor, Derek Ariss, without whose encouraging words I would have shied away from taking on such a project! Many thanks to my wonderful friends and colleagues in the CE ecosystem in India who have kept me learning constantly.

Abbreviations

AI	Artificial Intelligence
ALFs	Area-Level Federations
AMC	Ambikapur Mission City
APPLiA	Home Appliance Europe
BASF	Badische Anilin- und SodaFabrik (Baden Aniline and Soda Factory)
BBMP	Bruhat Bengaluru Mahanagar Palike
BEE	Bureau of Energy Efficiency
BGF	Blackrock Global Fund
BN	Banyan Nation
BRICS	Brazil, Russia, India, and China
CAGR	Compounded Annual Growth Rate
CAIF	Circular Apparel Innovation Factory
CBOs	Community-Based Organisations
CCOF	Circulate Capital Ocean Fund
CE	Circular Economy
CEA	Commissariat à l'énergie atomique et aux énergies alternatives (Alternative Energies and Atomic Energy Commission)
CEAP	Circular Economy Action Plan
CII	Confederation of Indian Industry
CNG	Compressed Natural Gas
CPCB	Central Pollution Control Board
CSR	Corporate Social Responsibility
DFC	U.S. International Development Finance Corporation
DRS	Deposit Refund Scheme
DWCC	Dry Waste Collection Center
EC	European Commission
ECERA	European Circular Economy Research Alliance
EDI	Electronic Data Interchange
EDIFACT	Electronic Data Interchange for Administration, Commerce and Transport
EDIFICE	Global Network for B2B Integration in High Tech Industries
EEE	Electronic and Electrical Equipment
EGD	European Green Deal

ELVs	End-of-Life Vehicles
EMF	Ellen MacArthur Foundation
ENEA	Agenzia nazionale per le nuove tecnologie, l'energia e lo sviluppo economico sostenibile (Italian National Agency for New Technologies, Energy and Sustainable Economic Development)
EPR	Extended Producer Responsibility
ESDM	Electronics System Design and Manufacturing
ESG	Environment, Social and Governance
ETF	Exchange Traded Fund
EU	European Union
EV	Electric Vehicle
E-waste	Electronic waste
FFG	Fashion for Good
FMCG	Fast Moving Consumer Goods
FY	Financial Year
G20 RE dialogue	Group of Twenty Resource Efficiency Dialogue
GACERE	Global Alliance on Circular Economy and Resource Efficiency
GDP	Gross Domestic Product
GHG	Green House Gas
GIZ	Deutsche Gesellschaft für Internationale Zusammenarbeit, GmbH
GPP	Green Public Procurement
GS1	Global Initiative on Developing Standards
GST	Goods and Services Tax
HDPE	High Density Polyethylene
I4R	Information for Recyclers
IETU	Institute for Ecology of Industrial Areas
INATBA	International Association for Trusted Blockchain Applications
INR	Indian National Rupee
InRP	Indian Resource Panel
IoT	Internet of Things
IP	Intellectual Property
IPP	India Plastics Pact
IUSCEI	Indxx US Circular Economy Index
IVL	Stiftelsen Institutet för Vatten- och Luftvårdsforskning
KRAs	Key Responsibility Areas
LDPE	Low-Density Polyethylene
LVP	Low-Value Flexible Plastics

MeitY	Ministry of Electronics and Information Technology
MLDL	Mahindra Lifespace Developers Limited
MLP	Multilayered Packaging
MMRPL	Mahindra MSTC Recycling Pvt. Limited (MMRPL)
MMT	Million Metric Tonnes
MNCs	Multinational Companies
MoEFCC	Ministry of Environment, Forests and Climate Change
MPCB	Maharashtra Pollution Control Board
MSTC	Metal Scrap Trade Corporation Limited
MWC	Mahindra World City
MWTESL	Mahindra Waste to Energy Solutions Limited
NDTV	New Delhi Television Ltd
NGOs	Nongovernmental Organisations
NITI Aayog	National Institution for Transforming India
OECD	Organisation for Economic Co-operation and Development
OEMs	Original Equipment Manufacturers
PCR	Post-consumer Recycled
PE	Polyethylene
PET	Polyethylene Terephthalate
PMC	Pune Municipal Corporation
PP	Polypropylene
PPEs	Personal Protective Equipment
PROs	Producer Responsibility Organisations
PSS	Product-Service Systems
PVC	Polyvinyl Chloride
QC	Quality Control
QR code	Quick Response code
R&D	Research and Development
RAAS	Resale-as-a-Service
RoHS	Restriction of Hazardous Substances
RTO	Research and Technology Organisation
SAMCLF	Swachh Ambikapur Mission City Level Federation
SCIP	Substances of Concern in Products
SCP	Sustainable Consumption and Production
SDGs	Sustainable Development Goals
SHGs	Self-Help Groups
SI	Social Innovation
SINTEF	Stiftelsen for industriell og teknisk forskning
SMEs	Small and Medium Enterprises
SPI	Sustainable Product Initiative
STEEP	Social, Technological, Economic, Environment. and Political
SUPs	Single-Use Plastics

SWM	Solid Waste Management
TMW	The Mahindra Way
TNO	Nederlandse Organisatie voor Toegepast Natuurwetenschappelijk Onderzoek (Netherlands Organisation for Applied Scientific Research)
TPP	The Protoprint Project
UKRI	UK Research and Innovation
ULBs	Urban Local Bodies
UNDP	United Nations Development Programme
USA	United States of America
USD	US Dollars
VAT	Value-Added Tax
VTT	VTT Technical Research Centre of Finland
WBCSD	World Business Council for Sustainable Development
WIEGO	Women in Informal Employment: Globalising and Organising
WRAP	Waste and Resources Action Programme
WWF India	World-Wide Fund for Nature – India
WWOW	World Without Waste

Enabling Framework for Circular Economy Transition and Policy Innovation: An Indian and Global Perspective

1

Ashish Chaturvedi, Dieter Mutz, Katharina Paterok, and Rachna Arora

Contents

DOI: 10.1201/9781003201816-1

1

Rising prosperity is both welcome and problematic. It is welcome in that it indicates substantial reductions in absolute poverty. It is problematic in that it increases pressures on our limited natural resources through unsustainable production and consumption patterns. The dominant linear economy model of make-use-throw has led to the unsustainable extraction of finite raw materials to meet the increasing demands. At the same time, such models contribute to rapidly rising quantities of waste, putting additional pressure on the local and global ecosystems.

At least two different sets of pathways may emerge from the pressures on ecosystems. On the one hand, such pressures could seriously hinder development outcomes and adversely affect the communities and countries dependent on natural resources. In such pathways, populations are dragged into conflict, misery, and poverty. On the other hand, such pressures also provide new opportunities and trigger transformational pathways which might lead to sustainable development. In such pathways, societies find means to confront the challenges, adopting more sustainable resource use to meet the rising aspirations. While the first set of pathways results from business as usual and inaction, the second set of pathways needs active engagement in transformational thinking and interaction between business, civil society, academia, and government.

The environmental movement has postulated the need for a change in balancing production of material goods, as evident in the Limits to Growth (Meadows et al. 1972). This need for change is now recognised more widely in debates on how to move from the throwaway to a circular economy (CE). According to a joint report of the Ellen MacArthur Foundation and the World Economic Forum,

> A circular economy is an industrial system that is restorative or regenerative by intention and design. It replaces the end-of-life concept with restoration, shifts towards the use of renewable energy, eliminates the use of toxic

chemicals, which impair reuse and return to the biosphere, and aims for the elimination of waste through the superior design of materials, products, systems, and business models.

(World Economic Forum and Ellen MacArthur Foundation 2014)

The ultimate aim is to have a circular system that eliminates waste and brings the embedded resources and energy in end-of-life products back into production and consumption processes through repair, reuse, and recycling. The circular economy is also taking an increasingly important role in the policy agenda. For instance, China passed a Circular Economy Promotion Law in 2008. In 2015, the European Union established a Circular Economy Strategy to move Europe into a more resource-efficient economy.

China made rapid strides, a decade before India, in this transformation to a circular economy. This transformation was partly driven by the realisation that it needs affordable sources of raw materials to meet the demands of its rapidly expanding industrial sector which serves as a global manufacturing hub. However, over the last few years, India has made rapid strides, driven by a concerted effort on waste management, aspirations for industrial sector expansion, the realisation of the growing need for raw materials, and the limits of the linear economy.

1.1 OUR POINT OF DEPARTURE

On August 15, 2021, Prime Minister Modi announced that 'The country is also emphasising on Mission Circular Economy. Our Vehicle Scrap Policy is a great example of the same' *(The Week* 2021). This statement is our point of departure for this book, with innovation at the centrepiece of the discourse. From being at the margins of discussion during the first decade of the twenty-first century, the circular economy is now at the forefront of discussion in India's highest echelons of policymaking. This journey from the margins to the mainstream is the focus of our interrogation.

However, it is critical to recognise that there is no blueprint that can be copied from one context to another to achieve the transformation from a linear to a circular economy. The experiences of the European Union (EU) and Japan might be instructive for the experience in countries like China and India. Still, the approaches and solutions would have to be uniquely country specific, but without neglecting – and not ignoring – the overall importance and relevance of global trade and international business cooperation. As we write this chapter, the narratives of decoupling growth from resource consumption and leapfrogging are being played out, not in text-books but in the global economy.

Situating such a large-scale and rapid change in the context of an economy would need country-specific analysis because multiple actors with divergent interests influence such transformative processes. These actors come from businesses, academia, government, and civil society. Also, they operate at different levels – local, national, and global. Some of these actors gain from the transformations, while others lose from them. Certain new actors emerge to take advantage of the opportunities provided by the transformations, while certain incumbent actors give way to these emerging actors. These multiple actors influence how the problem and its solutions are framed for their benefit. Such framings lead to the emergence of several narratives that often compete. At the same time, these actors, interests, and narratives interact with each other across levels and seek support from the forces that promote their interests across levels. For instance, the setting up of Extended Producer Responsibility (EPR)–led models in India requires the formal set-up for waste collection, dismantling, and recycling, which can potentially lead to diminished access to waste and subsequent loss of livelihood for informal actors, as their role is being eventually minimised. However, in the electronics waste stream in India, we have witnessed cooperation models set up between brand owners, Producer Responsibility Organisations (PROs), and social enterprises or informal sector workers directly to ensure better collection and recycling targets.

Our prime focus area is to examine the milestones and processes that have pushed the circular economy into the lexicon of policy and practice at the highest level in India. The interaction between domestic and international policies is the second focus of this chapter. Through a joint declaration signed in July 2020, resource efficiency (RE) and the CE are central aspects of the India–EU cooperation agenda. The narratives of circularity have burgeoned and prospered, not only in India and the EU, but in different parts of the world.

1.2 TURNING POINTS AND TRANSITIONS: FROM A LINEAR TO A CIRCULAR ECONOMY

In order to understand the present discussions and proposed strategies on how to move towards a full-fledged circular economy, one must understand the historical roots of our economic behaviour. In this section, we apply the Social, Technological, Economic, Environmental, and Political (STEEP) analysis as a

Social	Technology	Economic	Environment	Political
SDGs	Knowledge base in manufacturing	Industry Competitiveness	Sustainability	G 20 RE Dialogue
Informal sector	Disruptive & Innovative	Export restrictions	Consumption	BRICS
Jobs & skills	Digital	Import Dependency	Climate Change	GACERE
Consumerism	Localized solutions	Business Innovation	Material scarcity	EU-India Joint Declaration on RE and CE

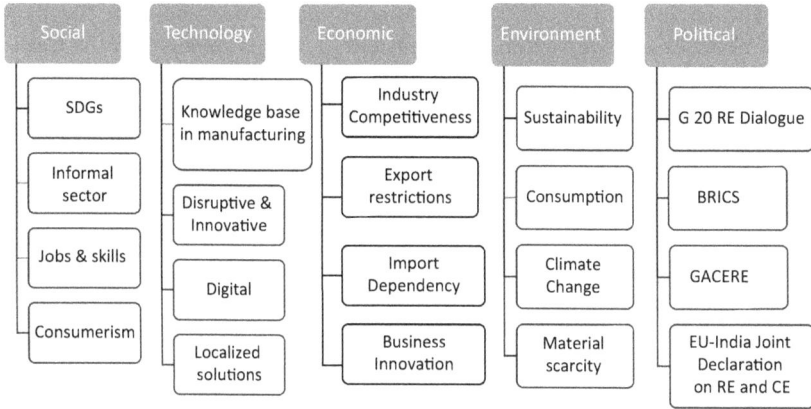

FIGURE 1.1   STEEP factors influencing CE transition and policy innovation in India

tool to evaluate different external factors and their impacts on the CE transition in India and policy innovation. We also identify the key milestones which brought forward circularity in India. In doing this, we also shed light on how the STEEP factors playing out in India were influenced by the discussions in and with the EU. Figure 1.1 outlines the STEEP factors influencing the CE transition and policy innovation in India.

1.2.1 Social

The United Nations 2030 Agenda for Sustainable Development and its Sustainable Development Goals (SDGs) encompass many elements of the circular economy. The principles of a circular economy are especially reflected in SDG 12, which focuses on sustainable consumption and production. India is committed to all the SDGs, and increasing circularity in the Indian economy plays an important role in realising India's sustainability goals over the next decades. Sectors such as mobility, agriculture, and construction, and the socioeconomic aspects connected to them, play a crucial role in the future growth of India.

The embedded sustainability in Indian ideology with a widespread social understanding of repair, reuse, recycling, and sufficiency of consumption remain extremely predominant in daily conversations and even in the scientific consultations. Even today, the inherent sustainability of traditional Indian culture remains a common refrain in conversations around sustainability. For several years, India maintained the 'Indian way,' with a slow but steady economic growth compared to Europe and other countries from the West.

The social changes in India's demographic landscape over the past three decades are substantial and continue to impact changes to the traditional Indian culture. India's middle class has doubled between 2001 (5.7%) and 2010 (12.8%) and continues to increase rapidly. This emergence of the new middle class has a substantial impact on people's behaviour and lifestyles. By 2030, India will move from being an economy led by the bottom of the pyramid to one led by the middle class. Nearly 80% of households in 2030 will be middle-income, up from about 50% today. The middle class will drive 75% of consumer spending in 2030 (World Economic Forum 2019). The increased spending power leads to changing trends in consumer behaviour, which, in a business-as-usual scenario of a linear economy, will lead to repercussions for the future of a multitude of societal needs and systems. Therefore, these changes in the demographic continue to drive an increased demand in circularity. Also, a systematic understanding of gender aspects in a circular economy transition will be key to inform behavioural preferences, the value of jobs in local value chains, and gender roles in imbibing sustainability (OECD 2020).

India's informal recycling sector is perhaps the largest agglomeration in an economy that is inherently based on the basic principles of a circular economy. A growing realisation and recognition by various actors in India of the role of the informal sector in scavenging materials from waste/resources have been a major step towards moving to an inclusive circular economy and just transition. Moving to formal repair, repurposing, and recycling systems as part of the circular transition is bound to enhance jobs and provide opportunities for upskilling. For instance, reuse, salvaging spare parts, product leasing, etc. would require manual operations and specialised skills for resource segregation (Circle Economy 2021).

1.2.2 Technological

Since the beginning of this century, many sectors in India, such as the automotive, pharmaceutical, and electronics sectors, have made advances in manufacturing. While they evolved to meet global benchmarks, these sectors created new opportunities for start-ups especially with the move towards global digitalised industrial development. The digital transformation that has taken place in recent years has reinforced India's position as a hub for digital technology and technological innovations. Sectors such as construction, food and agriculture, mobility, and vehicle manufacturing all serve as prominent sectors for disruptive and innovative digitalisation and technology applications, especially in the realm of start-ups but also for established businesses. Pairing circular economy aspects with digital technology has generated and continues to generate additional value creation.

Ongoing government initiatives like Digital India and India's investments in large-scale infrastructural projects, such as Swachh Bharat (Clean India) or the Smart Cities Mission, embrace the principles of circular economy as they aim to redesign sanitation and waste services and digitalise urban spaces to improve citizens' quality of life.

1.2.3 Economic

Despite the impressive growing and evolving economy – with annual Gross Domestic Product (GDP) growth rates of up to 6%–7.6% in 2018, leading to a rapid reduction in poverty – and the continuous growth of India's middle class, the challenges for Indian governments in the past three decades remain similar. Even today, the dependency on raw material imports, particularly critical metals; the dependency on power supply from nonrenewable energy sources; and the informal nature of many businesses still put pressure on resource consumption. Globalisation and the opening up of the borders to all kinds of consumer goods resulted in a slight erosion of the sustainability path of the past. The generation of increasing quantities of solid waste and waste water, significant air pollution, and inefficient land use are just a few of the negative symptoms of the 'new' economic growth which have threatened sustainable development in India.

Circular and resource-efficient approaches have the potential to improve resource availability, which is critical to the growth of industries and translates into reduced price volatility due to supply constraints or disruptions. By using resources more efficiently, or by utilising secondary resources, industries can improve competitiveness and profitability, since material cost is typically the largest cost for the manufacturing sector.

CE-based business innovations can also give industries an edge in the export market, as the experience of global leaders such as Germany and Japan have shown. One aspect which plays a crucial role here is that of export restrictions. Trade barriers, such as export restrictions, are frequently applied to secondary raw materials – largely scrap or electronic waste – in order to protect domestic industry interests. The consideration of internationally applicable norms is therefore an important element, as secondary raw materials play a significant role in the circular economy.

Finally, reduced import dependence for critical minerals, such as cobalt, nickel, and copper, helps to reduce the vulnerability of key industries to supply shocks. It further improves the country's trade balance and promotes economic stability (NITI Aayog 2017).

1.2.4 Environmental

The topic of sustainability and sustainable development has been gaining importance in India since the publication of the 12th Five-Year Plan (2012–2017) as it placed sustainable development centre stage. Till today, the debates on environmental regulations and what a sustainable economy could look like are underpinned by the understanding that sustainable development requires a systemic change which is enhanced by an enabling environment and policy framework.

Sustainable consumption and production patterns are a priority for the Government of India (GoI), which is reflected in various policies and programmes such as Make in India, Zero Effect – Zero Defect Scheme, Smart Cities, Swachh Bharat, or the Ganga Rejuvenation Mission. Sustainable consumption of resources is key for a rapidly developing country like India, which is dependent on increasing its overall resource consumption to meet its developmental goals. While the per capita consumption of materials in India is still low compared to the rest of the world, India's total resource consumption is quite high due to the large population size and is expected to increase rapidly. Between 1980 and 2009, India's total material consumption increased by 184%, making it the world's third largest consumer of materials, accounting for 7.1% of global material consumption. If current trends continue, India's material requirements are projected to be 15 billion tonnes by 2030 and 25 billion tonnes by 2050, with the biggest increases in the shares of fossil fuels, metals and minerals (Arora et al. 2016). This increased demand in combination with the already existing resource scarcity in certain sectors, such as the construction sector, imposes significant environmental burdens, many of which, particularly climate change, are becoming progressively more acute.

Resource extraction and utilisation are extremely energy intensive, thereby utilising a large amount of fossil fuels, which, even today, remain the main source of energy. This implies a strong correlation between resource use and greenhouse gas (GHG) emissions, which is a matter of urgent global concern. Energy-intensive industries with a high carbon intensity conflict with India's GHG reduction commitments under international climate-change agreements. India has taken steps to curb expected increases in GHG emissions, including launching efforts to increase the efficiency with which resources are used and by accelerating the adoption of clean technologies.

Thus, due to the dwindling resource availability, environmental destruction, and the challenges of climate change, developing sustainable models and learning from international best practices are a matter of great significance and urgency.

1.2.5 Political

The Agenda 2030 put forward an approach towards revisiting economic growth vis-à-vis environmental concerns and just transition. The Report of the Expert Group on Low Carbon Strategies for Inclusive Growth (2014) emphasised the need for India to introduce adequate and innovative tools (incentives, laws, market mechanisms). It also highlighted the need for a visionary approach for foresighted entrepreneurs (producers, dealers, recyclers) that support the required change towards a resource-efficient economy. The RE strategy (NITI Aayog 2017) elaborated the strategic actions to be undertaken by the GoI with key relevant ministries to steer proactive action across sectors for decoupling resource consumption from economic growth. As covered in Section 1.4, the G 20 RE dialogue provides a framework for prioritising domestic action and enhancing collaboration for knowledge exchange to move away from a linear economy.

1.3 RECENT DEVELOPMENTS IN MAINSTREAMING CE IN INDIA AND THE EU

Since the beginning of 2010, developments regarding India's and Europe's transition towards a circular economy continue to converge. The European Commission (2015) adopted the agenda on the circular economy focusing on resource-efficiency aspects with a strong focus on waste management. The European Commission has adopted a new Circular Economy Action Plan 2020 as one of the main blocks of the European Green Deal and Europe's new agenda for sustainable growth. It introduces legislative and non-legislative measures targeting areas where action at the EU level brings real added value. The European Green Deal (EGD), launched by the European Commission in December 2019, outlines a comprehensive framework of regulations and legislations.

At the same time, RE and CE as potential strategies for sustainable growth have increasingly gained traction in India. Figure 1.2 outlines the strategies, policies, and milestones in mainstreaming the RE and CE agenda in India. This is evident in the establishment of the Indian Resource Panel (InRP) at the Ministry of Environment, Forests and Climate Change (MoEFCC) with the aim of supporting the Government of India to elaborate an evidence-based policy on resource efficiency. The recommendations of the InRP were critical

FIGURE 1.2   Strategies, policies, and milestones in mainstreaming RE and CE agenda in India

in designing strategies such as the National Resource Efficiency Strategy, released by NITI Aayog in 2017. Multiple initiatives at the national level focus on a variety of circular economy and resource efficiency–related aspects, including national flagship programmes such as 'Make in India' or 'Clean India,' as well sectoral policies such as the national environmental policy, waste management rules, and research and development (R&D) for innovative technical solutions.

In 2019, MoEFCC released the draft national RE policy, which provides an overall framework for resource efficiency and secondary resource utilisation. The draft RE policy mentions the major policy goals and instruments needed for transitioning to a circular economy, viz. eco-design, sustainability standards, recycling targets, investments in jobs, technology access, and collaborative ecosystems as the major elements. This visionary draft policy brings to the forefront the policy goals as mentioned in the Ellen MacArthur Foundation's universal policy goals for transitioning to the circular economy (EMF 2021). Also, in 2019, NITI Aayog developed sector-specific RE strategies for four sectors, namely steel, aluminium, E-waste, and construction and demolition waste which cover a range of measures to reduce resource intensity in these sectors and waste streams.

1.4 ROLE OF INTERNATIONAL COOPERATION IN MAINSTREAMING RE AND CE

The system-level transformation that is needed to turn circular requires international partnerships, collaborations, policy alignment, and commitment (Ellen MacArthur Foundation 2021). The facilitation of dialogues between political partners is crucial to steer the developments and align the aforementioned often iterative policymaking when it comes to the circular economy. Globally, the G20 RE dialogue, which was initiated in 2017 in Germany, was a major kick starter of the international cooperation efforts towards mainstreaming RE and CE in a political context. The participation of India in the G20 and BRICS (Brazil, Russia, India, and China) summits led to commitments towards increasing cooperation across sectoral, priority-based topics across countries for knowledge exchange and innovative policy instruments. BRICS countries also set a precedent by creating a RE dialogue to create joint collaborative platforms and multi-stakeholder engagement for harmonised approaches and benchmarks. In 2021, India also agreed to be part of the Global Alliance of Circular Economy and Resource Efficiency (GACERE), to contribute to a global impetus for initiatives related to the circular economy transition, resource efficiency, and sustainable consumption and production.

The India–EU Resource Efficiency and Circular Economy Partnership, signed at the EU–India Summit in 2020, aims to strengthen and scale up EU–India cooperation in the areas of resource efficiency and the circular economy and promotes a convergence of positions at multilateral environmental negotiations, such as in the contexts of the G20 RE Dialogue or GACERE. Figure 1.3 showcases key elements of cooperation under the EU–India Joint Declaration on RE and CE, signed in July 2020. The facilitation of the exchange of information and expertise on issues related to CE and sustainable consumption and production on a global scale is especially relevant in response to the

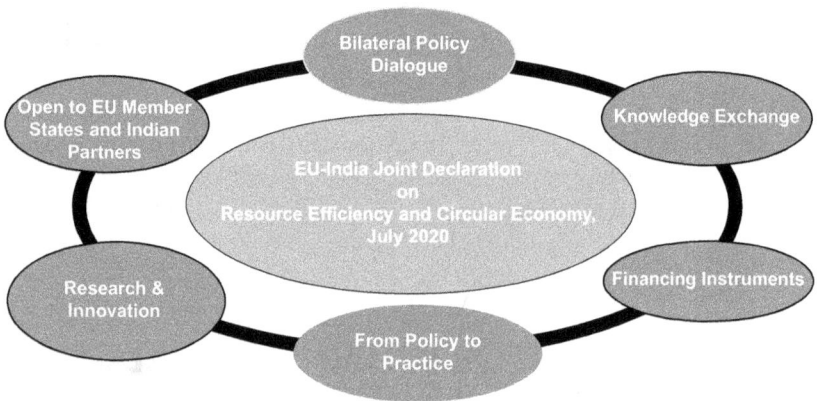

FIGURE 1.3   EU and India Joint Partnership on RE and CE

COVID-19 pandemic, when countries around the globe are working towards green recovery.

The New Delhi Statement on Environment during the 7th Meeting of BRICS Environment Ministers held in New Delhi, India in 2021 also high-lighted the need for the facilitation of knowledge exchange, sharing of best practices, and promotion of dialogues among the BRICS nations in the field of RE and CE (New Delhi Statement on Environment 2021).

1.5 POLICY INSTRUMENTS PROMOTING UPTAKE OF CIRCULAR ECONOMY

When assessing the macro narrative of India's developments regarding a transition towards a circular economy and how it gained momentum on a global stage, it becomes evident that it is very much an iterative and adaptive policy-making and governance process. India, just like the rest of the world, develops strategies on the go, as it aims to enhance its circularity in a volatile world which is characterised by rapid economic, environmental, and social change. This is also the reason for multiple revisions of existing rules over a short period of time. In the following section, a few policy instruments and enabling frameworks that promote the uptake of a circular economy are analysed.

To undertake the transformation to a circular economy, it is necessary to create an enabling environment and a supporting governance structure which consists of policies and measures based on efficiency, sufficiency, consistency, cooperation, and trust. The transition to a circular economy is promoted by a variety of policy instruments which – once introduced – ideally create an enabling framework and an ecosystem in which value chains can turn circular and resource efficient. Instruments of particular significance include EPR schemes, regulations and requirements for design for recycling and remanufacturing; Green Public Procurement (GPP); and the establishment of standards and certification schemes and corresponding monitoring frameworks.

Few countries around the globe currently support legislative approaches to increase circular and resource-efficient systems in a more comprehensive way. Figure 1.4 describes the policies and measures in place which look beyond waste management and address all stages of the lifecycle, especially product design and manufacturing processes, in the EU, Japan, India, and China.

While only a few countries take CE-related legislative action beyond waste management approaches and the promotion of waste prevention and recycling, sector-specific legislative action is increasingly being recognised as being key when creating an enabling environment for a circular economy. Based on this understanding, countries around the globe introduce policy instruments that address especially resource-intensive sectors with high environmental impacts and potential for secondary resource utilisation. India is currently developing a national mission on circular economy (NITI Aayog) covering 11 sectors ranging from scrap metal (ferrous and nonferrous) to used-oil handling (Sharma 2021). The action plans would focus on measures to design sustainable

EU's CE Framework	European Green Deal EU Circular Economy Action Plan Sustainable Product Initiative
Japan's CE Framework	Sound Material-Cycle Society Circular Economy Vision 2020
China's CE Framework	Circular Economy Promotion Law 2009
India's evolving CE framework	RE strategy 2017 Draft RE policy 2019 Sectoral strategies and action plans 2019

FIGURE 1.4 Global legislative framework and policies to address lifecycle thinking

products, enhance product longevity, and promote product take back through strict enforcement of policies and fiscal instruments to promote circularity.

However, we must understand that there is no single policy measure that, as a standalone, enables the transition to a circular economy. The various types of policy instruments – regulatory, economic, information-based, voluntary agreements, and behavioural tools – need to be combined in an effective mix to address the context with respective policy challenges and eventually lead to a system change.

1.6 EXEMPLARY POLICY INSTRUMENTS AND BEST PRACTICES PROMOTING CE

This section reviews selected policy instruments that promote circularity along the product lifecycle and highlights selected global best practices thereof. Transitioning to a CE requires substantial alterations to manufacturing industries, behavioural change at the consumer level, and adapted waste management practices. As per the recent report (Ellen MacArthur Foundation 2021), the universal policy goals that can support value-chain transition are institutional structures and identifying interconnectedness that influence reconsiderations of product design and usage; the promotion of product durability through repair; and innovation across value chains and the end-of-life stage of products. Some of these are discussed in the following section.

1.6.1 Eco-design

A product's lifetime and its environmental impact are determined by the design phase. Eco-design is a systematic and comprehensive approach for products aimed at reducing environmental pollution along the entire lifecycle through improved product design. It is based on an integrated lifecycle assessment. It therefore complements product requirements, such as functionality, safety, and the ratio of price and performance, with the requirement of environmental friendliness. Eco-design can be enabled by policies which support the design of high-quality products, which are durable, reusable, shareable, repairable, remanufactured, and recyclable. The elimination of waste and pollution lies at the core of the approach, while material loops are slowed and the need for circular business models increase.

Throughout the European Union, there is an increased interest in eco-design. With the Eco-Design Directive, the European Commission has for the first time created a framework for defining minimum requirements for the energy efficiency for specific product groups. The aim of the directive is to improve the environmental compatibility of energy-related products while taking into account the entire lifecycle by specifying design requirements. The directive is implemented through product-specific regulations that are directly applicable in all EU countries. It provides consistent EU-wide rules for improving the environmental performance of products, such as household appliances, information and communication technologies, or engineering. As part of the EU Circular Economy Action Plan, it was announced that an Eco-Design Directive will be implemented to ensure that devices are designed for energy efficiency and durability, reparability, upgradability maintenance, reuse, and recycling. This includes the introduction of a common charger requirement for electronic devices and EU-wide take-back schemes to return or sell back old mobile phones, tablets, and chargers. In addition to this, the use of hazardous substances in electrical and electronic equipment will be restricted.

In India, as per the MeitY's CE draft strategy, CE will necessitate a mix of policy approaches related to eco-design (like Restriction of Hazardous Substances [RoHS]) and sustainable manufacturing of electronics and electrical equipment (MeitY 2021). It will also focus on the development of CE business models, promoting subscription-based models, resource recovery through mining of precious resources, and the enhanced use of secondary resources.

1.6.2 Repair and Product Life Extension

An enabling environment for repair relies on a variety of factors, such as the access and the right to repair as well as a competitive market and the general mainstreaming of repair among consumers. Higher-quality products with increased durability increase the incentive for product repair and remanufacturing. Both activities are labour intensive, which would therefore additionally contribute to the creation of jobs and building up skills in the labour force.

Currently value chains in multiple sectors, from Electrical and Electronic Equipment (EEE) to textiles and furniture, are characterised by high consumption rates, short product lifecycles, and few repair options. The low quality of products leads to limited lifespans; this undermines the economic incentive for repairs, which end up being equally or more expensive than the purchase of a new product. Other barriers to repair include the risk of repair quality, the availability of spare parts, and the higher costs of repair for certain product categories.

Potential policies and initiatives that support the repair of products are manifold, such as eco-design requirements which lead to ease of repair, modular products, providing consumer information about product lifetime and reparability and take-back schemes, labelling schemes, and tax deductions for the repair sector and/or for consumers who choose to repair a product.

The European Commission's new Eco-Design Directive is the first legislation in Europe that includes the right to repair. The directive introduces reparability requirements for selected products, namely TVs, refrigerators, dishwashers, washing machines, and servers. A few European member states have also introduced tax-related measures to encourage citizens to get their products repaired. In 2017, the Swedish government introduced a tax reduction for repair and proposed new strategies to promote a market favoured by repair. With this, the government introduced tax reductions, with Value-Added Tax (VAT) being reduced to 12% from 25% for repair services, including, for example, repairs of shoes and bicycles. In addition, the Swedish government incentivises the repair of large household appliances by making 50% of the repair labour costs tax deductible up to a certain threshold. These tax incentives in Sweden aim to make repair more economically viable and accessible for the consumer (Riksdagen 2021).

As India underwent a second wave of COVID-19 in May 2019, M-19 initiative of Makers Asylum (an Indian start-up) initiated local and decentralised manufacturing of oxygen concentrators through open-source innovation (Makers Asylum 2021). However, post the second wave, thousands of concentrators became dysfunctional, so the team initiated the skilling of local communities to repair oxygen concentrators through a series of repair cafes to extend the life of the equipment (Makers Asylum 2021).

1.6.3 Extended Producer Responsibility (EPR)

Extended producer responsibility is a policy tool that assigns producers, manufacturers, and distributors responsibility for the end-of-life of their products. The responsibility can range from financial to operational responsibility, and it can affect product design as well as waste management patterns. The aim is to keep the environmental impact as low as possible throughout the entire product lifecycle, which is why the successful application of EPR schemes requires the manufacturer to know the material composition of their products to influence the product design.

EPR schemes are a cornerstone of a circular economy and contribute substantially to Germany's resource efficiency. The legal basis for EPR in Germany is laid down in the Circular Economy Act (*Kreislaufwirtschaftsgesetz*). This includes specifications for the development of durable products, the use of secondary raw material in production, and the return of products and environmentally friendly disposal after use. These objectives are supported by the prohibition of certain substances and mandating of labelling requirements and design of take-back schemes for manufacturers and retailers. Specific requirements for certain products are set by laws or regulations (e.g. for packaging deposit regulation), and in some laws and regulations, the scope of waste management is further specified, for example, for packaging, vehicles, batteries, and electrical and electronic equipment.

In 1991, the German government adopted the Packaging Ordinance (*Verpackungsverordnung*) in response to the ever-increasing volumes of packaging. With this first-of-its-kind ordinance, manufacturers and distributors were obliged to take responsibility for the packaging from production to environmentally friendly disposal. This was achieved through the determination of collection and recovery provisions. Dual systems for collection and disposal were subsequently introduced nationwide to uphold this responsibility. The German Packaging Ordinance gave rise to the implementation of national measures in neighbouring states such as Austria, the Netherlands, Belgium, and France, which in turn inspired the adoption of the European Directive 94/62/EC on Packaging and Packaging Waste of 20 December 1994, which is now legally binding for all EU member states.

India's e-waste (Management) Rules, 2016; the Plastic Waste Management Rules 2016; and the solid waste management rules all incorporate EPR-based legislation and targets for improved collection and recycling. With the MoEFCC having released an updated EPR notification under the Plastic Waste Management Rules in February 2022, EPR has been identified as one of the key elements enhancing a closed-loop economy by mandating use of recycled content, increasing recyclability as well as bringing traceability in the plastic-packaging sector.

1.7 CONCLUSIONS AND RECOMMENDATIONS

The circular economy and resource efficiency have risen in the policy priority in India over the last decades. Two overarching trends explain this phenomenon. The first set of trends are domestic. The growth in the middle class as well as technological advancements domestically are critical drivers for the emergence of the circular economy narrative in India. At the same time, there has been an increasing focus on the circular economy in economic and environment policy processes because of the need for sustained economic growth and poverty reduction balanced against environmental sustainability.

The second set of trends that have influenced the narratives of circularity in India are international developments. The circular economy has gained traction most prominently in the economies of the European Union as well as in comparable large economies such as China. The interconnectedness of the global economies as well as geostrategic considerations of resource security have strongly influenced an openness to policy innovations internationally. Moreover, the focus of long-standing partners to India, especially the European Union, on prioritising international cooperation on topics such as the circular economy has also led to knowledge exchange and capacity development.

These domestic and international trends do not work in isolation. The interaction of domestic and international drivers also has the potential for accelerating the adoption of a circular economy. Bilateral cooperation initiatives, which provide the platform for the interaction between the domestic and international trends, and cooperation (e.g. G20) are such potential accelerators. Several initiatives have emerged from the cooperation between India and the European Union and some of its member states (especially Germany). These examples suggest that fostering such cooperation and proactiveness on the part of the cooperating partners have the potential to create a global community of actors in the public and private sector that can push the transformation from a linear to a circular economy.

We believe that our analysis has the prospect to inform domestic as well as international policy and practice on the transformative potential of the circular economy. The STEEP analysis provides a stylised set of drivers that could elucidate in what manner a circular economy can gain traction in a developing economy and especially how these drivers have played out in the Indian context. Several recommendations emerge for policymakers from this analysis. First, a circular economy must be viewed within a broader lens of social, technological, economic, and environment policy rather than being restricted to a narrow environment policy lens. Recent policy developments in India, especially by the NITI Aayog, take an especially broad-based approach which is consistent with such an enabling ecosystem set-up.

Second, and relatedly, policy development on circular economy must take a whole-government approach rather than being siloed in line ministries. Third, while policy development must be a holistic government approach, the implementation of the policies has to take a whole-society approach. The critical role of the private sector in the transformation towards a circular economy must not be underestimated. However, the role of policy in pushing the private sector to adopt circular business models is crucial because it facilitates overcoming the inertia and comfort with the linear models. At the same time, in the Indian context, the private sector also must consider the critical role of the informal sector as a value-chain actor. The informal sector has the potential to play a transformative role given its experience and expertise in traditionally used circular business models.

The institutionalising of this approach might be realised by setting up an inter-ministerial and multi-stakeholder structure which takes up the work programme developed over the last decade in different line ministries. Such a structure could be developed along the lines of the Bureau of Energy Efficiency (BEE), a statutory body under the Ministry of Power of the Government of India. Our final recommendation would be to have a clear focus on a lifecycle approach rather than investing in piecemeal measures which might lead to burden shifting to other parts of the lifecycle. The CE-promoting best practices and instruments outlined above work best in conjunction with each other, rather than in isolation. If the transition to a circular economy has to be accelerated, it cannot be achieved by creating islands of excellence or small-scale initiatives. It must be implemented using a lifecycle approach by bringing the relevant stakeholders together and designing large-scale interventions.

REFERENCES

Arora, R; Banerjee, A. & Becker, U. 2016. *Material Consumption Patterns in India: A Baseline Study of the Automotive and Construction Sectors.* Deutsche Gesellschaft für Internationale Zusammenarbeit (GIZ) GmbH. GIZBaseline ReportSummary_SinglePages.pdf (international-climate-initiative.com).

BRICS India 2021. 2021. *New Delhi Statement on Environment.* doc202182731.pdf (pib.gov.in). Accessed on November 18, 2021.

Circle Economy. 2021. *Circular Jobs Definition Framework - Insights - Circle Economy.* Circle-economy.com. https://www.circle-economy.com/resources/circular-jobs-definition-framework. Accessed on September 20, 2021.

Ellen MacArthur Foundation. 2021. *Universal Circular Economy Policy Goals.* 1. [EN] Universal circular economy policy goals_Jan2021 (1).pdf. Accessed on October 5, 2021.

European Union. 2015. Circular Economy Action Plan. https://environment.ec.europa.eu/topics/circular-economy/first-circular-economy-action-plan_en. Accessed on January 12, 2022.

Maker's Asylum. 2021. M-19 O2 Oxygen Concentrator by Maker's Asylum (makersasylum.com). Accessed on November 10, 2021.

Maker's Asylum. 2021. Repair and Reuse – Fixing Oxygen Concentrators! - Maker's Asylum (makersasylum.com) Accessed on December 10, 2021.

Meadows Donella H; Meadows Dennis L; Jorgen R & Behrens III WW. 1972. The Limits To Growth. A Report For The Club Of Rome's Project On The Predicament Of Mankind. (donellameadows.org).

Ministry of Electronics and Information Technology (MeitY). 2021. *Circular Economy in Electronics and Electrical Sector.* https://www.meity.gov.in/writeread-data/files/Circular_Economy_EEE-MeitY-May2021-ver7.pdf. Accessed on November 2, 2021.

NITI Aayog. 2017. *Strategy Paper on Resource Efficiency.* Strategy Paper on Resource Efficiency.pdf (niti.gov.in). Accessed on November 10, 2021.

OECD. 2020. *Gender-specific Consumption Patterns, Behavioural Insights, and Circular Economy.* GFE-Gender-Issues-Note-Session-5.pdf (oecd.org). Accessed on November 10, 2021.

Riksdagen. 2021. Inkomstskattelag (1999:1229) Svensk författningssamling 1999:1999:1229 t.o.m. SFS 2021:1164 - Riksdagen. *Riksdagen.se.* https://www.riksdagen.se/sv/dokument-lagar/dokument/svensk-forfattningssamling/inkom-stskattelag-19991229_sfs-1999-1229. Accessed on November 1, 2021.

Sharma, Yogima. 2021. Fiscal Sops, Stricter Rules in the Works to Back Circular Economy. *The Economic Times.* https://economictimes.indiatimes.com/news/economy/policy/fiscal-sops-stricter-rules-in-the-works-to-back-circular-econ-omy/articleshow/87707093.cms. Accessed on November 10, 2021.

The Week. 2021. Highlights of PM Modi's Independence Day Speech - The Week. Accessed on September 20, 2021.

World Economic Forum. 2019. *Future of Consumption in Fast-Growth Consumer Markets: India.* World Economic Forum. https://www.weforum.org/reports/future-of-consumption-in-fast-growth-consumer-markets-india. Accessed on August 20, 2021.

World Economic Forum and Ellen MacArthur Foundation. 2014. *Towards the Circular Economy: Accelerating the Scale-up across Global Supply Chains.* https://www3.weforum.org/docs/WEF_ENV_TowardsCircularEconomy_Report_2014.pdf. Accessed on August 20, 2021.

Complexities and Challenges of the Circular Economy Transition

2

Patrick Schröder and
Manisha Anantharaman

Contents

DOI: 10.1201/9781003201816-2

2.1 SYSTEMS INNOVATIONS FOR CIRCULARITY: COMPLEXITIES AND CHALLENGES

The circular economy (CE) is a diverse paradigm offering both a vision and pathway for a global sustainability transition. Broadly, it calls for a move away from the current linear economic model of 'take–make–throw away,' in which resources are extracted, turned into products and services, consumed, and ultimately discarded (Anantharaman 2021). In contrast, in a circular economy, products and materials are kept in circulation for as long as possible. The circular economy concept applies lifecycle thinking and 'cradle to cradle' approaches that seek to use residues as 'food' for new products and processes. A shift to renewable energy sources is a basic requirement for the circular economy, alongside innovations in industrial sectors to preserve embodied resources.

To achieve a circular economy that is 'regenerative by design,' a shift from the current industrial growth society to a life-sustaining society of diverse regenerative cultures and systems that maintain the healthy functioning of ecosystems and the biosphere is necessary (Wahl 2021). The changes required to get to a circular economy will implicate a wide range of economic sectors, their supply chains, and consumers. Industrial and product design will need to change to make products more durable, reusable, repairable, and recyclable. Changes in product design will need to be anchored in new business models that in turn provide new technical and skilled-manual work opportunities. Given the interconnectedness of global value chains, achieving a system-wide transition will require unified quality standards for goods, services, and secondary raw materials, as well as new infrastructures, policies, and social norms to anchor and support new practices. In the medium to long term, these systems innovations will have wide-ranging implications for policy, global value chains, employment and skills, global trade, technology, and finance, and will require new forms of collaboration and close coordination among all stakeholders involved.

While the circular economy demands system-wide transformation, this transformation must start somewhere, and here innovation becomes key. Here, we identify five key areas for synergistic multilevel system innovation to support a just circular economy transition (see Figure 2.1).

(1) Innovations on the level of products and materials will entail new design approaches for products to be durable, repairable, modular, and biodegradable. New approaches such as design for repairability, modularity, and maintenance will be key to reconfiguring existing

Crosscutting digital innovations across sectors and value chains

E.g. Big Data analytics and artificial intelligence for reverse logistics systems; digitally enabled product-service systems, blockchain applications used in sustainable supply chain management

Bottom-up experimentation and innovation

Policy innovations

e.g. EU Circular Economy Action Plan, India's draft Resource Efficiency Policy; India's RE Strategy

Societal innovations

e.g. Behavioural changes enabling sustainable consumption through product-service systems and collaborative consumption

Business model and value chain innovations

e.g. new revenue schemes and customer interface such as pay-per-use models, reverse logistics and additive manufacturing

Innovation in materials and product design

e.g. Design for reuse & repairability, phase-out of hazardous chemicals from recycling loops

Top-down coordinated innovation processes

Macro level

Partnerships and inclusive processes for a just transition

Meso level

New technical and skilled-manual work opportunities

Micro level

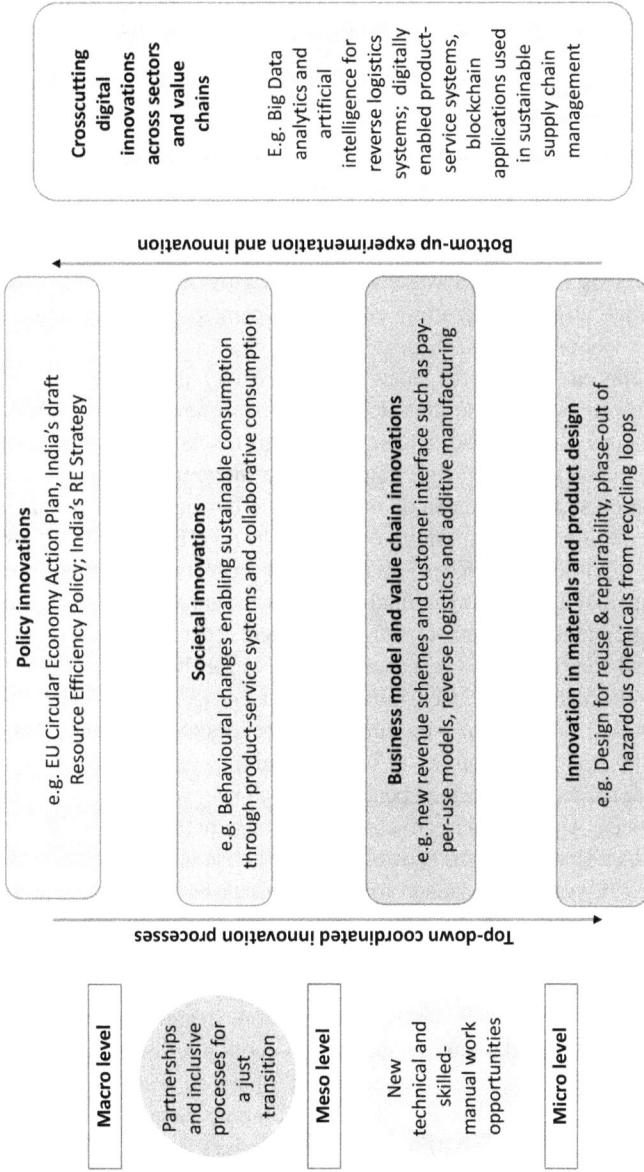

FIGURE 2.1 Multilevel innovation processes for circular economy transitions

models of planned obsolescence. Product redesign will also require the phasing out of toxic materials from loops for a clean and safe circular economy, particularly for recycling workers. New harmonised product information and reporting systems are needed together with digital tracing technologies that can track the composition of materials and eliminate harmful chemicals before they are reused and recycled (ChemSec 2021).

(2) Circular business model innovations aim to transform industrial value chains. One overarching concept driving business innovation is that of the performance economy, in which the producers of goods retain ownership over products and internalise all liabilities and costs of risk and waste associated with the end-of-life stages (Stahel 2010). Integrating circularity principles downstream (at the consumption and disposal stage) alters value capture and how benefits are generated. Innovations here include changes to ownership models and service delivery through new revenue schemes and customer interfaces such as pay-per-use models. Upstream (at design and manufacturing stage) circular interventions change value creation systems and can be implemented through digitally enabled reverse logistics and additive manufacturing (Pieroni, McAloone, and Pigosso 2019). These changes have implications for organisational and inter-organisational collaboration across value chains and implications for investors, consumers, and other stakeholders.

(3) Social innovation and behavioural changes for the circular economy include bottom-up citizen initiatives and engagement in local communities, alongside new consumption practices (Marchesi and Tweed 2021). A growing number of initiatives aim to generate low or zero-waste models with societal benefits through intensive use of goods and services. Also pertinent are changes to consumption patterns through product-service systems and collaborative consumption, such as shared mobility systems or laundromats, which can be more resource efficient while providing the same services. Grassroots initiatives like tool libraries and repair cafes are examples of social systems innovations that rely on citizen involvement and public support.

(4) The circular economy demands innovations in national policy, standards, and regulations on the macro-economic level. It is linked to national sustainable development plans as well as industrial strategies, such as the EU Circular Economy Action Plan, China's 14th Five-Year Plan (2021–2025), and India's Resource Efficiency Strategy. Specific policy tools include eco-design policies such as the European Union's new Sustainable Products Initiative or the circular economy ISO standards (59000 series by ISO TC 323) that

are currently under development. These new policies will drive not only product innovation, but also innovations in public procurement, infrastructure development, and financial products.

(5) Finally, digital innovations are cross-cutting and include new applications to increase resource recovery, transparency, and traceability to keep track of products, components, and materials. Big Data analytics and artificial intelligence can support reverse logistics systems and enable the shift to product-service systems, especially mobility services. Other examples of digital innovations are blockchain applications used in sustainable supply chain management (Saberi et al. 2019). These tools and data-driven metrics can enable businesses to understand the impact of their product components throughout the lifecycle, while also helping organisations develop clear targets to improve circularity over time.

When implemented synergistically, these diverse systems innovations for the circular economy transition will result in a profound systemic transformation of the way the world's economies function. While it is likely that it will generate a net-positive outcome in terms of employment opportunities, many workers, industries, and communities could also be adversely impacted. Transitions might also include trade-offs, and thus opposition. The predicted benefits of new business models, job creation, and reduction in waste and pollution for one country or demographic will, in many cases, result in adverse impacts on others. This is especially problematic if the transition harms already marginalised communities. For these reasons, applying a just transition approach is important for identifying which countries, sectors, communities, and workforces may be adversely affected by the various innovation processes. Identifying potential winners and losers through participatory 'road-mapping' can help shape effective cooperation mechanisms and partnerships nationally and internationally. It can also help develop policies and programmes to support those at risk of being left behind (Schröder 2020).

2.2 THE INDIAN AND SOUTH ASIAN CONTEXTS

As COVID-19 threatens to undo gains in the fight against extreme poverty, South Asia, as home to a majority of the world's poor, is facing significant economic challenges. At the same time, the intensifying climate and resource crisis has undercut the viability of resource-intensive development models. A just circular economy transition could enable South Asian countries to leverage the strengths of a large work force, pre-existing resource conservation practices, and vernacular knowledge to target development for poverty alleviation.

The regenerative elements of the circular economy can be found in rural India and across the wider South Asian region. Linear agricultural systems are high-input ventures and farmers are increasingly hard-pressed to sustain their livelihoods. In contrast, traditional regenerative farming systems that are focused on local communities, dietary diversity, and food security provide proven alternatives, as examples from rural communities in Rajasthan demonstrate (Schröder et al. 2019). These regenerative and restorative farming principles have benefits for topsoil regeneration, increasing biodiversity, and supporting the natural water cycle. These practices can be used for food crops as well as cotton and other fibres used in textile production. It supports business models that work on conserving natural resources such as soil, water, and biodiversity, as well ensuring fair compensation for smallholder farmers, who are the most vulnerable in the supply chains (Laudes Foundation 2020).

Looking at industrial sectors, it will be important for South Asian economies to not only keep up with the innovation trends in key manufacturing sectors such as textiles and garments, electronics, and plastics, but also to adapt them to local contexts. As a global manufacturing hub for textiles and garments, as well as the place where old clothes, electronics, and plastics sometimes go to die, South Asia will be critical to any global efforts at moving from a linear to a circular economy. A challenge for South Asian governments, businesses, and informal workers lies in maintaining and reasserting this centrality at a time when many EU countries and business are retrenching their supply chains within their borders as the circular economy becomes a green growth and job creation strategy in the EU.

The global textiles supply chain provides a good illustration of South Asia's centrality. India and Bangladesh are among the top five largest garment manufacturers in the world. In Bangladesh, the garment industry accounts for 15% of the country's GDP and 82% of total merchandise exports in Financial Year (FY) 2015, and it employs 4 million workers, of whom almost 80% are women (Asian Development Bank 2020). Cities like Panipat in India, the world's cast-off capital, are global hubs for textile recycling. Simply put, work and economic activities in South Asia bookend many of the global chains of production and consumption.

However, as labour rights and environmental activists have pointed out, while millions of people depend on jobs in these industries, textile production is a resource-intensive and polluting industry with poor work conditions. The total greenhouse gas emissions from textiles production, at 1.2 billion tonnes annually, are more than what is produced from all international flights and maritime shipping combined (Ellen MacArthur Foundation 2017). The garment industry has also been negatively impacted by COVID-19. Across the board, there is agreement that the time is ripe for reform, and any attempt to reform the global textiles value chain will most likely both start and end in South Asia. To make the sector circular, we still need to take into account the livelihoods of the people working in these industries.

In South Asia, where industrialisation is still in progress, innovation provides leapfrogging opportunities for sustainable industrialisation, sustainable lifestyles, and reducing environmental impacts such as air pollution or plastic waste (Schröder et al. 2019). India, in the last couple of decades, leapfrogged from an agrarian to IT service-based economy where it has opportunities to reinvent the manufacturing sector with circularity embedded into the system.

Circular innovation could increase the competitiveness of businesses and industries both nationally and internationally. Electronics and automobiles are two sectors in India where circular economy principles, especially reuse and remanufacturing approaches, appear in India's Production Linked Incentive Schemes. India's National Policy on Electronics has the aim to position India as a global hub for Electronics System Design and Manufacturing (ESDM) by creating an enabling environment for the industry.

Circular product specifications have the potential to disrupt existing value chains and trade – with potential negative impacts on suppliers and small and medium enterprises (SMEs) in low- and middle-income countries. For instance, new product innovations for circularity and regulations that demand the adoption of eco-design criteria will apply to all suppliers across the value chain irrespective of geography (Barrie and Schröder 2021). New policy developments such as the EU's Circular Electronics Initiative will be relevant for India. To meet new eco-design criteria, suppliers might have to invest in upgrading knowledge, technology, and skills. Considering these linkages between emerging innovations in downstream consumer product markets and upstream manufacturing that supply large consumer markets will be necessary (Smart Prosperity Institute 2021).

The textile sector here presents another useful example. Bangladesh's export-oriented textile sector is highly vulnerable to changes in demand patterns (as seen during the pandemic). Manufacturers and suppliers need to be proactive and anticipate changes in consumer demand, global supply chains, and trade patterns. If brands based in Europe and the US move towards integrating more recycled textiles in their supply chains, ensuring that manufacturers and suppliers to the brands invest in the technology, expertise, and skills to work with these materials would be key. This presents an opportunity to integrate circular textile innovations into the production processes to reduce waste, save resources, and generate higher value products.

Circular innovation trends also have impacts on employment. While the total number of jobs could increase in a circular economy transition, the geographical distribution of new jobs is potentially highly uneven, with developed countries winning high-value jobs in repair, remanufacturing, and recycling to the detriment of these jobs in South Asian countries (Barrie and Schröder 2021). Thus, it will be important to identify sectors where India and other South Asian countries can shape or even lead these innovation trends towards circularity. This needs to be complemented by investments in repair, remanufacturing, and recycling infrastructures, as well as relevant skills training to support the growth of these sectors for a domestic circular economy industrial base and boosting local employment.

2.3 CE TRANSITION CHALLENGES: STRENGTHS, OPPORTUNITIES, AND EMERGING SOLUTIONS IN INDIA AND SOUTH ASIA

An important area of distinction and potential competitive advantage in a circular economy is its already-existing informal circular economies. South Asia's informal economy is perhaps the largest agglomeration of already-existing circular economies. Millions of people in India and South Asia make their living by extracting value from waste. These circular actors range from waste pickers and gleaners who work in municipal dumps to extract reusable and recyclable materials from trash to community micro-enterprises and large factories that aggregate, dismantle, repair, and refurbish everything from discarded electronics to old jeans. The formal economy relies on and appropriates the value produced by the informal economy. Waste pickers reduce the cost of public waste-management services by diverting recyclables away from landfills, reducing waste transportation costs. Recycling also produces environmental value (Tucker and Anantharaman 2020; Dias 2012).

More than 80% of all South Asia's workers engage in informal activities, and over 90% of the region's businesses are informal. This informal economy has been devastated by COVID-19, with the World Bank documenting an increase in both extreme and mid-level poverty (Bussolo and Sharma 2020). The pandemic is exacerbating existing vulnerabilities and precarity in informal work. The lack of employment protection and social security renders informal workers vulnerable to market fluctuations (Tucker and Anantharaman 2020). Working with discards, especially in the context of limited access to technology, protective equipment, or workplace protections, comes with a range of health costs, often disproportionately borne by women, scheduled castes, children, and religious minorities (Dias 2016). Wastes from Europe and North America are often shipped to countries like India, China, and Malaysia where cheap labour can process it (Gregson and Crang 2015).

A major opportunity and challenge for a just circular economy transition is exploring how to leverage the knowledge, skill sets, and networks of informal workers while improving work conditions and incomes. Alongside improving social conditions, investing in the capacity and skills of the informal economy is required to increase the value realised from waste, to shift away from downcycling of engineered plastic components or 'over-repair' of inefficient equipment and machines, for example, old cars with high emissions contributing to urban air pollution.

In the following sections, we highlight some examples of innovative ideas and business models in a number of sectors including plastic packaging, textiles and garments, and digital sharing platforms that could serve as inspiration for further action towards circularity.

2.3.1 Digital and Technological Innovations to Increase Resource Recovery, Traceability, and Materials Quality

At the product level, large consumer brands are beginning to redesign their packaging to reduce waste. For example, Tata Chemicals uses 4000 tons of multilayer plastic film every year for packing its consumer products in retail pouches. The multilayer plastic packaging used currently is difficult to recycle, so the company's new adhesive laminate was launched as a monomer material that is PE (polyethylene) as opposed to two different substrates PET-PE (Polyethylene terephthalate-Polyethylene). Recycling processes will, in theory, be simplified as operators have to deal with only one material. However, challenges with the scale-up of these new packaging innovations include compatibility issues with existing machinery, the need for investments in infrastructure for collection and sorting, and the required changes in consumer behaviour (Tata Sustainability Group n.d.).

In addition to larger brands, the past years have seen a proliferation of newer companies that deploy digital and technology innovations to increase resource recovery, traceability, and materials quality. Today, there is a diversity of start-ups and established companies that use mobile and digital technology to innovate waste recovery. Digital technologies are playing an increasingly critical role in complex systems like urban waste management, especially in the presence of a large informal economy, which would have been otherwise difficult to integrate into circular supply chains. Digital innovations that make it easier for recyclers to collect postconsumer waste ease transaction costs.

Mindtree, a technology company, hosts a platform called I Got Garbage that aims to integrate stakeholders along the informal–formal waste value chain. Through I Got Garbage, bulk generators of dry waste such as apartment complexes can request waste pick-up from informal waste pickers. The waste generator can either donate or sell the waste to the informal collector, with rates negotiated by the intermediary Mindtree platform. The platform also requires digital payments and invoicing, which brings previously untaxed cash flows into the realm of regulatory and tax oversight.

The current processes by which plastic waste is collected, cleaned, and processed for remanufacturing by small-scale recyclers neither produce plastics that are considered 'human-contact' safe nor are they recovered in a closed-loop system. This is because of a consistent lack of investment in informal recycling processes.

Looking to fill this gap, Banyan Nation, one of India's first vertically integrated plastic recycling companies, has developed a proprietary plastic cleaning technology to convert postconsumer and postindustrial plastic waste into high-quality recycled granules. Banyan Nation's social innovation lies in the ways in which it has integrated thousands of informal recyclers into its supply chain through a data intelligence platform.

While these digital and technological innovations can help integrate informal waste economies into global recycled value chains, additional policy and business model innovation is required to make sure that these partnerships improve incomes and livelihoods for informal actors. Here, new forms of incorporation and certification could enable private sector actors to prioritise environmental and social metrics above profit, and convene ethical collaborations with informal recycling communities. Intellectual and financial investment in informal enterprises could help ensure that existing enterprises and workers are not displaced by newcomers, and have the opportunity to benefit from job creation in the circular economy.

2.3.2 Inclusive Business Models and Circular Market Innovations

South Asian circular economy business models have the opportunity to integrate social inclusiveness approaches such as base-of-pyramid solutions and participatory approaches. One example to consider here is with respect to fair trade models. Currently, one of the main factors hampering waste recovery and recycling is that recycled plastics in many cases are more expensive than virgin plastics.

Plastics For Change, a social enterprise based in Bengaluru, has introduced the novel idea of fair-trade recycled plastics to address this issue. It works with waste pickers and scrap dealers as aggregators, promising a better price for their plastics in exchange for an assurance of safe working conditions and no employment of child labour. Companies looking to maintain their reputations as good ecological stewards, along with consumers looking to play a part in addressing the ocean plastics crisis, are willing to pay more for this value-added, high-quality recycled plastic.

Certifying products through Fair Trade or other labels can also be a step to incentivise the uptake of alternative packaging and alternatives to single-use plastics. Bags made from widely grown crops like jute and sisal in Bangladesh, or the use of banana and palm leaves for serving food, are already prevalent in South Asia. Certifications for jute, sisal, or other biodegradable products could spur more investment and innovation in alternative packaging. Finding ways to add value to these products and expand their use beyond small-scale, local, or traditional markets could create jobs in small and medium industries, while also moving South Asia away from the use of more environmentally damaging single-use options. There is also potential to export these to markets in Europe and beyond.

Most circular business model innovations currently focus on superior customer value, resource efficiency, and longevity of products by closing or slowing material loops. However, to create social value, they also need to focus on ethical sourcing, quality of work, social relevance, and contribution to well-being. An example of this approach in India is the work being done by the Circular Apparel Innovation Factory (CAIF) on seeding and scaling circular innovations for the textile industry. When the COVID pandemic hit in 2020, CAIF began to include a strong component of inclusiveness into their work with the Circular Plus project (Nancy Charaya 2021).

Not all circular business model innovations have a social focus, but many have the potential to contribute to improving access to services with lower resource use and environmental impacts. Examples of this include furniture and electronics rental through digital platforms such as Rentomojo in all major Indian cities. Large consumer-facing brands are experimenting with circularity. For example, in July 2021, Hindustan Unilever Limited launched an in-store vending machine in a Mumbai mall for its home care products, which will enable customers to refill plastic bottles for home products (Livemint 2021). These pilot initiatives are still in the early stages and will require scaling up to achieve significant shifts in the market.

2.3.3 Policy Innovations in India and South Asia: From Extended Producer Responsibility to a Comprehensive Resource Policy

Several South Asian countries have created policies to deal with waste in a manner that reduces landfilling and incineration. These policies are a strong first step towards a circular economy transition; however, the next generation of policymaking needs to shift focus from end-of-life guidelines to a comprehensive resource policy that connects production and consumption. A comprehensive resource policy framework will also include upstream interventions, including circular design to address planned obsolescence, right-to-repair legislation, bans on the production and retail of certain single-use items, and fiscal policy measures. The upcoming regulations in Europe under the Circular Economy Action Plan, including the Sustainable Product Initiative (SPI), single-use plastics ban, and other related policies, are an indication of how policy frameworks will drive circular innovations on different levels – from products to business models and across value chains.

In the past decade, the Indian government, responding to pressure and advocacy by environmental organisations, has revised its municipal, plastic, and e-waste management rules to emphasise some zero-waste practices such as source segregation, recycling, composting, and bioenergy. Parallel efforts at waste prevention and management have been pursued by national governments in Bhutan and Pakistan. These policies are focused on reducing the amount of uncollected waste and diverting waste from landfills while also enabling some resource recovery. Alongside waste-management rules, plastic bag bans are in place in several Indian states as well as in Bangladesh, Sri Lanka, Nepal, Bhutan, and Pakistan. While these bans only cover a subset of hard-to-recycle, low-quality plastics, they are a step in the right direction. However, implementation and effectiveness of these bans is varied.

Alongside policy innovations that decriminalise waste picking and facilitate waste picker access to waste, South Asian countries need to place more focus on developing and strengthening Extended Producer Responsibility (EPR) principles and guidelines. EPR is a principle that seeks to hold the producers of waste accountable and responsible for the end-of-life of their products. It shifts some of the costs of waste management from the public to private actors, and in the process can drive innovation to reduce waste production through upstream changes in how things are designed and manufactured.

At the time of writing, India has published draft guidelines on EPR plastic waste regulations that could serve as a model for India and the South Asia regions. India's E-waste (Management) Rules came into effect in 2016, and one of its highlights is the guidelines for EPR. However, currently the guidelines offered on EPR suffer from several flaws related to the lack of capacity of implementing organisations, and they lack a clear understanding of the diverse waste and material streams and the nature of informal work.

EPR guidelines for plastics need to be tailored to the diversity of materials that make up waste streams, as well as the diverse economic arrangements that exist to manage them in the informal economy. There are pre-existing informal value chains for some plastics like PET and HDPE (High Density Polyethylene). In those situations, EPR's focus should be on strengthening these value chains through investment in capacity building and physical infrastructure. For materials like multilayered plastics that currently have low or no value and are therefore not collected by informal actors, the approach has to be different. Creating artificial markets through mandatory buy-back programs and spurring innovation to find ways to biochemically reprocess these materials can help enable the cost-effective recovery of these plastics, which are otherwise a major source of marine plastic pollution.

EPR policy in South Asia needs to be created with full participation from informal workers with multistakeholder participatory platforms created at the city, state, and national levels (Chandran et al. 2019). Underpinning these efforts should be a commitment to fair remuneration alongside zero-waste principles.

An example of how national industrial policy can promote both circularity and social protection can be seen in Bangladesh's Ship Recycling Act, updated in 2018, which stipulates that ships must be recycled only in specific zones and that employers must provide workers with life insurance and other social protection.

2.4 CONCLUSIONS: CE LESSONS FROM INDIA AND SOUTH ASIA FOR THE REST OF THE WORLD

In this chapter, we argue that the transition to a just and inclusive circular economy will require synergistic, systemic innovation across economic sectors and supply chains. India and South Asia's experiences combining social and policy innovation to upgrade existing circular practices in situ while creating new pathways and possibilities could provide inspiration for a global joint vision. To ensure that these transition processes do not worsen social inequities internationally and intranationally, we highlight the importance of reforms to financing models, international coordination, and the need to move beyond market-based mechanisms.

Reforms to financing will be vital to harnessing the knowledge and capacities of informal workers to maximise value, while also improving livelihood conditions. Lack of access to finance and capital for small and medium enterprises, particularly in the informal sector, limits efforts to upgrade technology or improve work conditions. Impact investing in informal enterprises can help enhance access to capital to upgrade existing processes. However, there is also an important role here for philanthropic capital. Furthermore, there is a need for public financing and blended finance models that increase investor confidence and draw in private capital.

In the coming years, we can expect eco-design product standards to move beyond energy efficiency and climate-related impacts to also include circularity criteria. Most consumer products are designed in Western countries, and those designs often do not match with the recycling infrastructure available in South Asia. Therefore, greater coordination between designers, manufacturers, users, and recyclers will be essential to develop useful and implementable standards.

While market-based mechanisms are necessary, they are not in themselves sufficient to secure a socially just circular economy transition. Government leadership and coordination will be crucial for circular businesses to flourish, especially the manufacturing value chains. For innovative financing, trade, and business models to deliver social and environmental benefits to the most marginalised in South Asia, it is important to get the goals right. To this end, we emphasise the importance of public policy and investment, created with the full participation of informal workers and other key stakeholders.

In conclusion, circular economy innovations are an opportunity for South Asia to recognise the value of livelihoods and of safe and healthy cities beyond narrow economic metrics of profit or growth. Key to an inclusive and just transition to a circular economy will be non-competitive collaboration between countries, industry, and community organisations, motivated by a shared goal to achieve sustainable wellbeing.

REFERENCES

Anantharaman, Manisha. 2021. 'Reclaiming the Circular Economy.' *The Oxford Handbook of Comparative Environmental Politics*. July 14. doi:10.1093/oxf ordhb/9780197515037.013.30.

Asian Development Bank. 2020. *Asian Development Outlook (ADO) 2020: What Drives Innovation in Asia?* Asian Development Bank. https://www.adb.org/ publications/asian-development-outlook-2020-innovation-asia. Accessed 9 Jan 2022.

Barrie, Jack, and Patrick Schröder. 2021. 'How to Use Global Trade for a Just Circular Transition.' *CircularEconomy Earth*. August 25. https://circulareconomy.earth /publications/how-to-use-global-trade-for-a-just-circular-transition. Accessed 9 Jan 2022.

Bussolo, Maurizio, and Siddharth Sharma. 2020. 'COVID-19 Has Worsened the Woes of South Asia's Informal Sector.' *World Bank: End Poverty in South Asia*. December 7. https://blogs.worldbank.org/endpovertyinsouthasia/covid-19-has -worsened-woes-south-asias-informal-sector. Accessed 9 Jan 2022.

Chandran, Pinky, Kabir Arora, Marwan Abubaker, and Nalini Shekar. 2019. 'Valuing Urban Waste: The Need for Comprehensive Material Recovery and Recycling Policy.' *Hasiru Dala.*

Charaya, Nancy. 2021. 'The Circular+ Project Updates: Collective Effort and the Pathway to Circular Fashion.' *CAIF Blog.* February 16. https://circularapparel.co/blog/2021/02/16/the-circular-project-updates-collective-effort-and-the-pathway-to-circular-fashion/. Accessed 9 Jan 2022.

ChemSec. 2021. 'Circular Economy.' https://chemsec.org/policy-and-positions/circular-economy/. Accessed 9 Jan 2022.

Dias, Sonia Maria. 2012. 'Not to Be Taken for Granted: What Informal Waste Pickers Offer the Urban Economy.' *The Global Urbanist.* http://globalurbanist.com/2012/11/27/waste-pickers. Accessed 9 Jan 2022.

———. 2016. 'Waste Pickers and Cities.' *Environment and Urbanization* 28 (2). SAGE Publications Sage UK: London, England: 375–90.

Ellen MacArthur Foundation. 2017. 'A New Textiles Economy: Redesigning Fashion's Future.' https://ellenmacarthurfoundation.org/a-new-textiles-economy. Accessed 9 Jan 2022.

Gregson, Nicky, and Mike Crang. 2015. 'From Waste to Resource: The Trade in Wastes and Global Recycling Economies.' *Annual Review of Environment and Resources* 40. Annual Reviews: 151–76.

Laudes Foundation. 2020. 'Regenerative Production Landscape.' October 14. https://www.laudesfoundation.org/latest/press/2020/laudes-foundation-idh-and-wwf-india-come-together-to-power-a-regenerative-production-landscape. Accessed 9 Jan 2022.

Livemint. 2021. 'HUL Sets up Vending Machine to Lower Plastic Waste.' *Mint*, July 9, sec. Companies. https://www.livemint.com/companies/news/hul-sets-up-vending-machine-to-lower-plastic-waste-11625811338028.html. Accessed 9 Jan 2022.

Marchesi, Marianna, and Chris Tweed. 2021. 'Social Innovation for a Circular Economy in Social Housing.' *Sustainable Cities and Society* 71 (August): 102925. doi:10.1016/j.scs.2021.102925.

Pieroni, Marina, Tim McAloone, and Daniela Pigosso. 2019. 'Business Model Innovation for Circular Economy: Integrating Literature and Practice into a Conceptual Process Model.' *Proceedings of the Design Society: International Conference on Engineering Design* 1 (July): 2517–26. doi:10.1017/dsi.2019.258.

Saberi, Sara, Mahtab Kouhizadeh, Joseph Sarkis, and Lejia Shen. 2019. 'Blockchain Technology and Its Relationships to Sustainable Supply Chain Management.' *International Journal of Production Research* 57 (7). Taylor & Francis: 2117–35. doi:10.1080/00207543.2018.1533261.

Schröder, Patrick. 2020. *Promoting a Just Transition to an Inclusive Circular Economy.* London: Chatham House. https://www.chathamhouse.org/2020/04/promoting-just-transition-inclusive-circular-economy. Accessed 9 Jan 2022.

Schröder, Patrick, Manisha Anantharaman, Kartika Anggraeni, and Timothy J. Foxon. 2019. *The Circular Economy and the Global South: Sustainable Lifestyles and Green Industrial Development.* London: Routledge.

Smart Prosperity Institute. 2021. 'Primary Materials in the Emerging Circular Economy: Implications for Upstream Resource Producers and Primary Material Exporters.' https://institute.smartprosperity.ca/PrimaryMaterialsCE. Accessed 9 Jan 2022.

Stahel, Walter R. 2010. *The Performance Economy*. 2nd edition. Basingstoke: Palgrave Macmillan.

Tata Sustainability Group. n.d. 'Closing the Loop: The Circular Economy in Action.' https://www.tatasustainability.com/pdfs/Resources/Closing_the_Loop_-_10_ initiatives_of_Circular_Economy_at_the_Tata_group.pdf. Accessed 9 Jan 2022.

Tucker, Jennifer L., and Manisha Anantharaman. 2020. 'Informal Work and Sustainable Cities: From Formalization to Reparation.' *One Earth* 3 (3). Elsevier: 290–99.

Wahl, Daniel Christian. 2021. 'Circular Economies & Regenerative Cultures.' *Age of Awareness*. June 14. https://medium.com/age-of-awareness/circular-economies -regenerative-cultures-f1bff04895c8. Accessed 9 Jan 2022.

Financing for a Circular Economy

3

Mayuri Wijayasundara, Srikanth Prakash, and Pavithra Mohanraj[1]

Contents

Financing plays the most critical role in enabling the transition from a linear to a circular economy (CE). This is not only because the capital infusion needed to fuel such a transition is significant but also because financing for technology, infrastructure, and monitoring is critical to scale up business models that can create a transformative and regenerative shift to create a circular closed-loop economy. Yet, the methods to finance the CE transition are in their infancy and need much innovation as well as discussion.

DOI: 10.1201/9781003201816-3

As an alternative economic concept, the circular economy embeds several principles that tie the economic and environmental performance of activities together. For the circular economy to replace the current economic system, its acceptance and adoption by multiple sets of stakeholders is vital. While environmental considerations may have an important part in this transition, its success in terms of long-term adoption may be based on its ability to bring economic returns to all stakeholders. In this context, the financing mechanisms that drive the transition and the fundamentals of financing associated with the new economic system are two main areas that need to be critically looked at for a major transformation.

This chapter discusses

1. how the concepts of economic exchange, financial consideration, and ownership are redefined in a circular economy and how they result in the emergence of product-service systems;
2. how financing plays a crucial role in funding initiatives during the transition and establishment of a new economic model.

3.1 REDEFINING ECONOMIC EXCHANGE IN TRADITIONAL TRANSACTIONS AND EMERGENCE OF PRODUCT-SERVICE SYSTEMS

In traditional selling of products or services, economic exchange mostly occurs only at the time of product or service purchase, where the consideration in exchange typically is 'currency' for 'products or services.' In a circular economy, the economic exchange shifts to a setting with more touchpoints between manufacturers, sellers, and users. In this setting, products, components, and materials are encouraged to be in circulation for a longer period, providing the maximum utility value, mostly through the involvement of multiple stakeholders. These stakeholders can include repairers, remanufacturers, or users. Also, the use or utility of a product is encouraged to shift beyond one use cycle by a single user who purchases the product. Thus, the direct exchange of ownership for a transactional value is not the only manner in which purchasing happens in a circular economy, with several other forms of economic exchange emerging as CE models develop. In the section below, we discuss economic exchange in the context of product-service systems.

3.1.1 Redefining the Fundamentals of Utility

The value a consumer places on the experience of a product, which we typically call 'utility,' plays an important role in defining the economic exchange in a circular economy. In attempting to keep products, components, and materials in use while providing the highest utility and value, the access to the product is disassociated from the ownership of the product. The value paid is in return for gaining access to the product, rather than to gain ownership of the product. This disassociation provides a fundamental basis to initiate discussion around selling utility services instead of products. The key questions to consider in developing product-service systems include:

- What do consumers need products for? This prompts us to understand the fundamental human need associated with consumption. (e.g., a washing machine is needed to wash the dirty clothes or maintain the cleanliness of the clothes).
- What needs to be performed to fulfil the customer's need? This prompts us to understand what needs to be done and what needs to be achieved (e.g., activity-wise: having a device and being able to wash the clothes; performance-wise: having a wardrobe of clean clothes).
- What is the consideration paid? This prompts us to define the service created to fulfil the customer need associated with a product, which then determines the economic exchange in consideration.

3.1.2 Description of Product-Service Systems

In the alternative economic model of a circular economy, there is an increased emphasis placed on product-service systems, challenging the manner in which transactions typically take place whereby products are exchanged against the currency. Product-service systems (PSS) can generally be defined as 'an integrated bundle of products and services which aims at creating customer utility and generating value' (Boehm and Thomas 2013), where the service offering of these can be presented in a variety of ways, such as:

1. Result-oriented: The user is sold access to the result that a product generates.
2. Utility-oriented: The user is sold access to the product for a number of cycles of use.

3. Term-oriented: The user is sold access to the product over its length of use.
4. Product-oriented: The user is sold ownership of the product along with services to maintain the product during its period of use (e.g., access to repair, faulty product replacement).

Table 3.1 describes how different PSS can be classified based on product ownership and service offered, and illustrates these PSS with the example of illumination provided by a light bulb.

With each of these options, the buyer and the seller/manufacturer exercise varying degrees of control and bear different types of risks over the same asset

TABLE 3.1   Classification of Product-Service Systems

TYPE OF PRODUCT-SERVICE SYSTEM	OWNERSHIP OF THE PRODUCT/COMPONENT/MATERIAL	ASPECT SOLD AS THE SERVICE	ASPECT SOLD AS THE SERVICE (USING THE EXAMPLE OF ILLUMINATION PROVIDED BY A LIGHT BULB)
Result-oriented	Relevant manufacturer or seller at the customer interface	Performance provided by the product	Area illuminated
Utility-oriented	Relevant manufacturer or seller at the customer interface	Number of cycles of use of a product	Number of hours or switch on–off cycles used
Term-oriented	Relevant manufacturer or seller at the customer interface	Length of use of a product	Number of months hired
Product-oriented	Buyer	Repair services, faulty product replacements, warranty	Faulty bulb replacements and warranty

(or product) that provides a service to a consumer. The case study below examines how the application of these concepts is developing in India.

CASE STUDY 1: PRODUCT-SERVICE SYSTEM TO LEASE FURNITURE: FURLENCO – FURNITURE FOR RENT

Case observations: The business model adopted by Furlenco leases furniture and home décor to customers. Furlenco adopts a full-stack model whereby design and manufacturing is carried out in-house, enabling Furlenco to provide trusted in-house product maintenance and repurposing services. The company has raised a total of US$36 million in funding as of May 2020 (Kapoor and Vij 2021).

Case analysis: The success of a rental model like Furlenco's relies on the customer's perception of value in having the flexibility to modify their home décor by swapping the leased products. The flexibility offered by Furlenco's model would especially appeal to a highly mobile urban population, including the young working demographic. In such a model, the service provider needs to match the customised requirements of multiple customers and constantly re-model the products to meet the latest trends in the interior designing space.

Financing this type of model would require additional capital to establish reverse logistics and refurbishment services, along with traditional product development resources. Accounting methods will also have to be reconsidered for this model as the products will continue to be owned by Furlenco and these assets will keep generating recurring revenue for the business. This will also have a bearing on how such new models are financed. Another key aspect that will affect the success of such a model will be the real-world rates of use and return for the individual products.

If scaled up, businesses like Furlenco could reduce the material demand and ecological footprint of the furniture industry significantly by repurposing the same furniture for use by multiple users and extending the life of the product beyond a single user. However, such a business model will also be exposed to risks associated with customer accountability for the misuse of and damage to the leased product. Similar business models are emerging in other sectors. Some examples include leasing clothing and fashion accessories (Flyrobe), chemical leasing, jewellery hiring, drinking-water vending systems that can be purchased on a pay-per-litre model, etc.

3.2 THE CRUCIAL ROLE OF FINANCE IN TRANSITIONING TO AND ESTABLISHING A CIRCULAR ECONOMY

With such fundamental changes in the economic system associated with a circular economy (as discussed in the previous section), the business and financial sectors need rethinking and evolution. With the product-service system distinctively identifying and disassociating access and ownership of products, there is a need to redefine several financial principles. These could potentially address assets and liabilities and economic exchange to account for differences in stewardship, custodianship, and ownership of assets (which include inventory, consisting of products, components, and materials) towards a circular economy. With that, there is a need to redefine elements such as collateral which are important in financial transactions.

The Circular Economy Finance Guidelines published by ABN AMRO, ING, and Rabobank represent a step towards creating uniform guidelines for recognising and defining relevant elements to be considered for financing CE projects (ING 2018). The guidelines focus on how investments are allocated and managed, and the method of assessment and selection of projects. It also provides information on how to classify an equity or debt product as a circular financing initiative.

Some key aspects to note with respect to financing initiatives related to the circular economy are:

3.2.1 Financing Collaboration

The transition to a circular economy requires the collaboration of multiple value chain partners, both in the upstream and the downstream areas. Equally, the level of reliance on other parties to initiate and effectively complete this transition is relatively high compared to other kinds of sustainability initiatives which can be carried out within the boundaries of an organisation. Therefore, financing the transition to a circular economy would require funding complex network models of organisations in order to realise the full impact and outcomes of initiatives and projects.

3.2.2 Financing Business Transformation

For businesses, the intensity of transformation towards a circular economy will be higher compared to the typical slow transition towards producing general sustainability outcomes. Not only will the transformation have to be rapid, but the spectrum of activities involved will also be broader, as the main trade activities need to shift to a different model. The opportunities for value capture in circular economy models will also be significantly different.

For businesses generally, transformation or transition to a new system becomes justifiable through an increase in financial gains or lowering of long-term business risk.

Typically, there is a perception that initiatives towards transition to a circular economy involve an additional cost to business. However, businesses tend to underestimate the potential for lowering long-term business risk by moving away from the linear economic model that inherently involves significant risks.

The lowered risk can be explained from multiple perspectives, such as:

1. reducing the dependency on overseas supplies and external markets by encouraging more local exchange of nutrients and services;
2. gaining assurance as to the availability of resources as the reliance on extracted resources decreases, while that on recycled materials or regenerative resources increase;
3. gaining assurance as to continued demand in export and global markets, where different emerging consumption models (such as PSS) continue to enable regenerative consumption.

Financial institutions such as banks are increasingly diversifying their investment portfolios by investing in sustainable or green investments as a means of reducing portfolio risk. Corporates or the public also have the option to invest in these through the stock market. There is an increased emphasis placed by investors on the Environment, Social, and Governance (ESG) performance of businesses in qualifying for financing. Increasingly, the aspect of improved reputation, mainly on account of not being perceived as 'purely-profit-oriented' entities (Ozili 2021), has become important for businesses to remain attractive for a wide range of investment options.

3.2.3 Financing Circular Economy–Related Initiatives through Investments

Currently, the annual investment in the linear economy is about US$35 trillion, and investment in CE attracts only 2% of the total amount invested in linear economic business models. Funding received for CE in different sectors such as the fashion, automotive, and agricultural sectors were 5%, 6%, and 0.1% of the total investment made in the respective sectors (Circular Online 2021).

Banks, institutional investors, and insurance companies can play a key role in helping raise capital for companies operating circular business models via venture capital funding and investing through equity and debt markets. An alternative approach is to directly finance projects through private and blended finance initiatives. The banking sector can offer to raise finance and develop relevant financial instruments for such innovative projects. Insurance companies can build their expertise in emerging CE models and develop tailor-made insurance schemes for PSS-based products or services, thereby minimising risk of exposure for circular businesses.

Globally, there are several capital investments that have been made to facilitate the transition towards a circular economy. Some examples are as follows:

1. Blackrock Global Fund (BGF) – Circular Economy: This is an equity fund managed by the world's largest asset manager, BlackRock, Inc. The fund invests in equity securities of companies contributing to accelerating the transition towards a circular economy. The total capital of the fund amounts to about US$ 2 billion, which was invested in businesses adopting and enabling CE (Blackrock 2021).

2. Direxion WWOW: This is an exchange-traded fund (ETF) managed by Direxion, a US-based provider of leveraged ETFs. The World Without Waste (WWOW) ETF fund provides exposure to the Indxx US Circular Economy Index (IUSCEI). The IUSCEI tracks the performance of 50 companies leading the transition towards a circular economy, which operate under five sub-themes (Neuwirth 2021):

   a. Sustainability of resources
   b. Resource recovery
   c. Lifecycle extension
   d. Sharing platforms
   e. Product as a service

3. Alphabet Corporate Bond: The corporate bond issued by Alphabet (US$ 5.75 billion) in 2020 is the largest corporate sustainability bond in history, with proceeds to be used for a number of focus areas including circular economy and design. Committed to maximising the reuse of resources across its supply chains by facilitating circular design, the company has used recycled plastics in millions of its products, including Google Nest which is being made with 100% recycled plastic (Holbrook 2020).

4. Impact investing: Impact investment firms such as Aavishkaar Capital invest in enterprises and projects which create a positive social and environmental impact along with generating a financial

return. Aavishkaar Capital has invested in companies like NEPRA Resource Management Private Limited in India, which are at the forefront of circular economy solutions in India (Roy 2020).

3.2.4 Facilitating the Financing of Local and Micro Activities

A circular economy promotes local value chains and local circulation of both technical and biological nutrients, as keeping most economic activity within a single region provides better control for undertaking regenerative design, which can have a greater impact. Microbusinesses that collect byproduct and waste streams for processing, conduct repair services (at the doorstep or within the local vicinity), or facilitate the selling of second-hand and used items are highly encouraged in a circular economy. Supporting micro-industries which need financial and technical assistance to scale up their business is also essential to achieve CE. With these sectors remaining informal and often unregulated, the access to finance remains a challenge, and this will need to be resolved through increased support, formalisation, and facilitation through social enterprises or public sector involvement.

3.2.5 Blended Finance Approaches

In order to address some of the key barriers faced by private entities that intend to invest in developing nations, two key focus areas are:

- to reduce the real and apparent risk in CE-based projects;
- to improve returns against the risks inherent in CE projects as compared to traditional investments.

In this context, blended finance is being increasingly advocated to fund CE projects, especially in developing countries (Merchant 2021, Raes 2021).

Two major platforms facilitating blended finance are Sustainable Development Investment Partnership and Convergence. Both platforms attempt to increase the use of public or philanthropic funds as catalytic capital to improve the risk/return profile of the transaction in order to attract private investment.

In the past decade, India has attracted blended financing in the energy, agriculture, and finance sectors, amongst others (Convergence 2017). The case study described here illustrates the need for and the use of blended financing to support emerging circular economy ventures in India.

CASE STUDY 2: CIRCULATE CAPITAL CASE STUDY

This case study was developed based on interviews with representatives from Circulate Capital.

Case observation: Circulate Capital is an investment management firm aiming to address the issue of ocean plastic pollution in South and Southeast Asia. The firm's goal is to catalyse capital to develop circular plastic value chains, where plastic waste is recovered and turned into a new, reusable resource for further use.

The firm established the Circulate Capital Ocean Fund (CCOF), a USD 106 million fund, via a blended financing model bringing together public and private sector players. Eight major private sector players including Procter & Gamble, Dow, Danone, Chanel, Unilever, The Coca-Cola Company, and Chevron Phillips Chemical Company LLC have committed capital to the fund. The fund is also backed by the U.S. International Development Finance Corporation (DFC) in collaboration with USAID, which provides a 50% guarantee on selected loans up to $35 million, thereby de-risking the CE financing initiatives from the private sector (HT Brand Studio 2020). Circulate Capital has currently invested USD 39 million in six different companies in India that are capable of recovering and recycling plastic and packaging waste and that generate a market through repurposing the waste.

Case Analysis

The fund is focused on addressing the lack of investments in areas of growth, expansion, and acquisition stages of circular economy innovations and infrastructure in the South and South-East Asian region, with the Indian waste management ecosystem to receive half of the total capital raised in the CCOF. The firm understands that limited investments are available for relatively early-stage Indian start-ups supporting the transition to a circular plastics economy. For these businesses to scale and have a wide-reaching impact on the plastic problem in India, institutional investors are essential. However, most institutional investors have stayed away from financing ventures in this space due to a 'missing middle' of investible entities. The 'missing middle' can be attributed to the nascent stage of the plastic waste innovation ecosystem and the resulting lack of a robust pipeline of investible opportunities, the lack of a proven track record of companies, and a lack of active funds in this space. Circulate Capital attempts to alter this mindset and catalyse capital to demonstrate the ability of such investments to generate economic and environmental gains (Circulate Capital 2021).

The investment strategy adopted by Circulate Capital is based on four key pillars:

- **Focusing on investments rather than grants** – This approach ensures that the capital infused into a business would result in a level of return that proves the model as well as having quantifiable socioeconomic and environmental impact, and that can generate accountability within beneficiaries.
- **Systemic change** – The capital needed is amalgamated from multiple stakeholders such as Fast-Moving Consumer Goods (FMCG) businesses, chemical industries, and retail businesses with a keen focus on improving recycling practices. Such investors are also key stakeholders as they already incorporate or plan to incorporate the CE-based solutions from the investee companies into their supply chain. Diversifying the investor pool across the plastics value chain is key to effecting systemic change, which cannot be accomplished by any single entity or investor.
- **Scalability and replicability** – Investments will be assessed and shared to ensure scalability and replicability.
- **Unlocking co-investment** – Collaborating with private investors to co-invest in selected projects would provide additional capital to such projects, with the successful track record of the Fund becoming a critical lever to bring in additional investors into future portfolios (HT Brand Studio 2020).

The six Indian businesses that attracted Circulate Capital investments are small and medium enterprises (SMEs) pioneering solutions in waste management and plastic recycling value chains. The investments from CCOF into these companies help address key issues in the waste recovery industry, such as fragmentation, lack of traceability, and low quality of recycled materials, along with generating safe, stable, and dignified employment. CCOF's support is also critical to these companies in dealing with the impacts of the COVID-19 pandemic as they are uniquely positioned to revolutionise the solutions necessary to build back stronger. These six companies fall under three innovation strategies to scale resource recovery (HT Brand Studio 2020):

1. Scaling upcycling to add value to waste
   a. Srichakra Polyplast Private Limited and Dalmia Polypro Industries Private Limited have expertise in developing food-grade bottles through bottle-to-bottle recycling technology.
   b. Deeya Panels Private Limited (Ricron) has pioneered a solution to recover and upcycle low-value multilayer plastic waste (one

of the largest contributors to ocean plastics) into high-quality building materials.

   c. Lucro Plastecycle Private Limited has developed a closed-loop material recovery process through which flexible plastic waste is recycled into new high-grade flexible films and shrink wraps.

2. Scaling digitisation

   a. Rapidue Technologies Private Limited (Recykal) developed India's first 'waste-commerce' business, creating multiple digital products to improve efficiency, transparency, and traceability in waste material transactions. Some of their solutions include Recykal Marketplace, a digital platform connecting bulk waste generators, aggregators, and recyclers; and Extended Producer Responsibility (EPR) Loop, a digital exchange for brands to connect with Producer Responsibility Organisations (PROs) and recyclers to fulfil EPR obligations.

3. Scaling waste collection and sorting in collaboration with cities

   a. NEPRA Resources Management Private Limited sources dry waste from marginalised waste collectors in the informal sector at stable and fair costs, and processes this waste into a quality recyclable form which is then supplied to recyclers.

The CCOF is structured as a strategic fund consisting of corporate limited partners since the value offered to the contributors of the fund is in the form of developing sustainable feedstock for their businesses. The CCOF also has a technical advisory committee with representation from investors' research and development (R&D), procurement, and corporate sustainability teams. Some of the key reasons which attracted Circulate Capital to India are government initiatives like the Swachh Bharat Mission and a high number of promising start-ups led by aspiring Indian entrepreneurs who are not reluctant to establish manufacturing businesses or to transform their family business towards developing CE-based solutions (Infinitive 2020).

3.3 CONCLUSIONS

In India's transition towards a circular economy, financing plays a key role, both in terms of providing incentives for the parties involved to change their current practice and creating new and sustainable operating models for organisations.

The general observation, as applicable to any context, is that the emergence, development, and sustenance of new mechanisms of economic exchange and financing for a circular economy depends on specific circumstances. These are related to industrial and technological development as well as region-specific socioeconomic conditions, although some fundamental principles apply universally.

This chapter has identified some of these basic principles associated with the financing of a circular economy and showcased case studies and examples from India, providing insights into how the evolution is likely to take place in this area to support India's transition to a circular economy. PSS models, such as Furlenco's, enable the circulation of materials through technical loops, but fundamentally alter the economic exchange of 'currency' for 'product.' Furlenco's model is based on continual revenue generation from the provision of rental services as opposed to sale of the product itself at the time of purchase, with several implications for financing such models. Different CE financing models that facilitate collaboration and transformation to CE, as well as impact-based investments, are discussed in the chapter. The chapter also discusses blended finance instruments which are needed to fast-track the transition to CE given the nascency of many CE models in India. The crucial role played by blended finance in facilitating a CE transition in India is also elaborated with a case study on Circulate Capital's investments in India.

NOTE

1. Data collection for development of the case study on Circulate Capital was supported by Pavithra Mohanraj.

REFERENCES

Blackrock. 2021. *BGF Circular Economy*. Blackrock, Accessed on July 25, 2021, https://www.blackrock.com/uk/individual/products/317822/blackrock-circular-economy.

Boehm, Matthias, and Oliver Thomas. 2013. "Looking beyond the rim of one's teacup: A multidisciplinary literature review of product-service systems in information systems, business management, and engineering & design." *Journal of Cleaner Production* 51:245–260, July 15, 2013, https://www.sciencedirect.com/science/article/abs/pii/S0959652613000243.

Circular Online. 2021. "$1.3tr spent on circular economy each year, but 'more needed' to reap full investment rewards." *Circular Online*, July 19, 2021 https://www.circularonline.co.uk/news/1-3tr-spent-on-circular-economy-each-year-but-more-needed-to-reap-full-investment-rewards/.

Circulate Capital. 2021. "Our investment portfolio." *Circulate Capital*, Accessed on July 25, 2021, https://www.circulatecapital.com/investments.

Convergence. 2017. "Blended finance-regions." *Convergence*, Accessed on July 25, 2021, https://www.convergence.finance/blended-finance#regions.

Holbrook, Emily. 2020. "Google issues largest corporate sustainability bond of any company in history." *Environmental Leader*, August 5, 2020, https://www.environmentalleader.com/2020/08/google-issues-largest-corporate-sustainability-bond-of-any-company-in-history/.

HT Brand Studio. 2020. "Circulate Capital invests to scale India's circular economy for plastic waste." *Mint*, December 10, 2020, https://www.livemint.com/brand-post/circulate-capital-invests-to-scale-india-s-circular-economy-for-plastic-waste-11607595018682.html.

Infinitive. 2020. "Financing circular innovation - Circulate capital", Innov8ive!: A deep dive into innovating for a circular economy." December 22, 2020, https://podcasts.google.com/feed/aHR0cHM6Ly9mZWVkLnBvZGJlYW4uY29tL2luZmluaXRpdmUvZmVlZC54bWw/episode/aW5maW5pdGl2ZS5wb2RiZWFuLmNvbS9lZWFuLmNvbVmFuLmNvbS9MzcyMzRiYYS00NGQ5LTNkOGUtODAzYy0zNTg1NjU5NDE4YTQ?sa=X&ved=0CAUQkfYCahgKEwiInYa9kKX1AhUAAAAAHQAAAAQmwE.

ING. 2018. "ABN AMRO, ING and Rabobank launch finance guidelines for circular economy." *ING Bank N.V.*, Last Modified 9 July 2018. https://www.ing.com/Newsroom/News/ABN-AMRO-ING-and-Rabobank-launch-finance-guidelines-for-circular-economy.htm.

Kapoor, Anuj Pal, and Madhu Vij. 2021. "Want it, rent it: Exploring attributes leading to conversion for online furniture rental platforms." *Journal of Theoretical and Applied Electronic Commerce Research* 16 (2):188–207. https://doi.org/10.4067/S0718-18762021000200113.

Merchant, Aakif. 2021. "How blended finance can support India's COVID-19 recovery." *Convergence*, July 21, 2021, https://www.convergence.finance/news-and-events/news/7rj5FQPcHS8F2LIdapZztd/view.

Neuwirth, Aaron. 2021. "Direxion launches world without waste ETF, 'WWOW'." *ETF Trends*, January 14, 2021, https://www.etftrends.com/direxion-launches-world-without-waste-etf-wwow/.

OECD. 2021. "Material consumption (indicator)." Organisation for Economic Co-operation and Development, Accessed July 25, 2021, https://data.oecd.org/materials/material-consumption.htm.

Ozili, Peterson K. 2021. "Circular economy, banks, and other financial institutions: What's in it for them?" *Circular Economy and Sustainability*: 1–12, April 30, 2021, https://mpra.ub.uni-muenchen.de/107397/.

Raes, Patrick Schröder. Jan. 2021. "Financing an inclusive circular economy." Chatham House, July 19, 2021, https://www.chathamhouse.org/2021/07/financing-inclusive-circular-economy/03-de-risking-financing-circular-economy-0.

Roy. 2020. "Waste management firm Nepra raises series C funding from Aavishkaar, circulate capital." *VC Circle*, November 13, 2020, https://www.vccircle.com/waste-management-firm-nepra-raises-series-c-funding-from-aavishkaar-circulate-capital.

Social Innovation for the Circular Economy: An Example of Informal Waste Recycling

4

Swati Singh Sambyal

Contents

DOI: 10.1201/9781003201816-4

The implementation of circular economy (CE) in resource management has largely focused on technological solutions with less or no emphasis on changes in user behaviour and social practices. Full implementation of a CE in cities requires a combination of changes in technologies and resource management infrastructure as well as changes in production and consumption practices. It is acknowledged that the CE in cities can be fully implemented only if institution-driven initiatives (top-down) are combined with society-driven actions (bottom-up). However, it is an irony that bottom-up initiatives in cities are rarely investigated as they are harder to identify and scale up. This is largely due to the fact that they are mainly managed by small locally-based groups of people, and are limited by regulatory, political, and infrastructural constraints (Marchesi and Tweed, 2021).

4.1 THE INFORMAL SECTOR IS CRITICAL FOR THE CIRCULAR ECONOMY

Informal waste recycling sector is the real game-changer in the journey towards a circular economy in India. The overall data regarding how much waste is being managed and recycled by the informal sector in India is very poor. About 1% of the urban population, or more than 15 million people, earn their living informally in the waste sector (World Bank. 2018). Employment in India is overwhelmingly informal, with over 90% of all workers employed under informal arrangements. In urban areas and in Delhi, a significant share is informally employed (around 80%). Nationally, 92% of women and 90% of men are informally employed (WEIGO. 2020). There are about 2.1 million waste pickers in the country, with 1.5 million in urban India alone. As per the India Plastic Recycling Report, 2019, there are 1–1.5 million informal workers directly involved in the recycling business.

The informal sector comprises of individuals, families, and/or micro-enterprises having unorganized activities which are neither taxed nor reported on by authorized entities. Under Solid Waste Management (SWM) services, they operate as waste-pickers at primary collection centres (for instance dhalaos) or at dumpsites, as itinerant waste buyers (kabadiwalas), small junk-shop dealers and medium-big waste godown owners.

Waste pickers usually rank lowest in the social stratum of informal urban occupations, with a vast majority being women and children. Many are also migrants, unskilled, illiterate, or part of the lowest caste hierarchy in India and are unable to find other employment opportunities (Shravanthi Kanekal. 2019).

Most of the informal actors in the waste management sector face a severe lack of identity, recognition, acknowledgement, appreciation, and basic human rights. Most do not have any social security as they do not have bank accounts and are forced to depend upon unpredictable cash flows. They are not made aware of the best and advanced practices of waste management, and there are no compliances with minimum wage, working hours, health, safety and environment standards, and other basic rules and regulations (Saahas Zero Waste. 2019).

4.2 ROLE OF THE INFORMAL SECTOR IN THE RECYCLING ECONOMY

About 6.02 million metric tonnes (MMT) of plastic gets recycled yearly in India (PlastIndia Foundation. 2019). This is about 70% of the total plastic waste that is generated in the country. There are over 100 organised recycling units (42 in PET recycling) and 10,000 plus unorganised recycling units (refer Box 4.1).

BOX 4.1   RECYCLING OR DOWNCYCLING?

In India, recycling majorly involves 'downcycling' into lower-grade products. The downcycled products cannot be used for food-grade packaging as there are no standards for recycled products and they are mostly used in lower scale applications. Waste plastics cannot be recycled beyond 3–4 times; since it degrades with each recycling. The recycling processes usually undertaken involve sorting of plastic waste (based on colour, transparency, hardness, density, and opacity of the scrap) which is then sent to granulators to obtain granules which are then fed to the converters to make finished plastic products.

Most of such units (granulators and convertors) function as single machine extruding units using technologies that are local, outdated and polluting, and are often located in low-income group areas such as slums or unauthorised colonies. Plastic scrap storage is done in the backyards, and washing is done in open drums – often termed as backyard recycling.

Low value plastic items (carry bags, gutkha/ paan masala sachets, styrofoam, disposable cutlery, wrappers of packaged food) form a major part of the plastic leakages in Indian cities. Presently, only three types of thermoplastics are recycled – high-density polyethylene for 20% (HDPE), polyethylene terephthalate for over 80% (PET), and polyvinyl chloride for 45% (PVC). The categories polypropylene for 7.6% (PP), polystyrene for 2.4% (PS), and low-density polyethylene for 25% (LDPE), although partially recyclable, are generally not, due to the economic unviability. (Swati Singh Sambyal 2019)

In India, 60% of plastics (such as polyethylene terephthalate [PET] bottles, high density polyethylene [HDPE] containers, polyvinyl chloride [PVC] pipes) are recycled. However, these are mostly downcycled in an informal setting (The Energy and Resources Institute. 2021). There is still a lack of structured integration of the informal sector into formal waste management systems despite the Solid Waste Management Rules 2016 mandating it.

For collection, the recycling industry depends heavily on the informal sector, such as waste pickers and kabadiwalas (small or itinerant buyers) (Figure 4.1). The informal workers channelise the items to small/intermediate traders/ aggregators, from where they are sent to apex traders and finally, to the recycling units (*see Figure 4.2: Informal sector in recycling value chain in India*). Different categories of recyclable dry waste are picked up and sorted by the informal workforce and then channelised (refer Figure 4.3).

FIGURE 4.1   A woman worker from the informal sector. Photo credits: by Swati Singh Sambal, April 2021.

Items such as PET bottles and plastic containers can fetch a price of 15–20 Indian Rupees (INR) per kilogram (kg). On average, a high-quality bucket made of virgin plastic can fetch INR 25–40 per kg. Items like carrier bags and wrappers of packaged foods (chocolates, chips, or sweets) do not get collected as they are usually highly contaminated and fetch lower prices, between 2 and

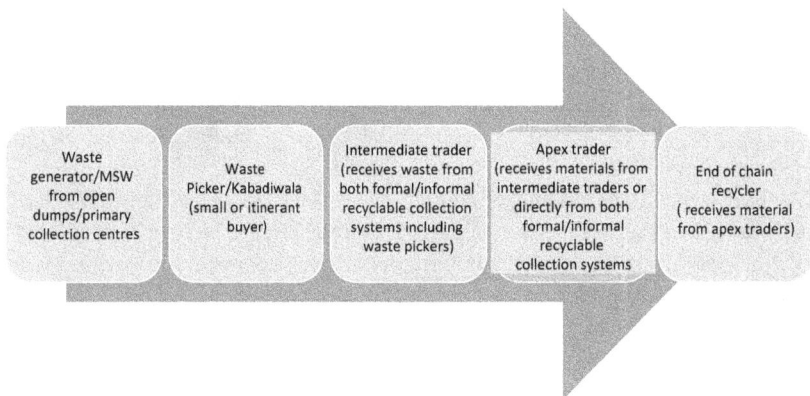

| Waste generator/MSW from open dumps/primary collection centres | Waste Picker/Kabadiwala (small or itinerant buyer) | Intermediate trader (receives waste from both formal/informal recyclable collection systems including waste pickers) | Apex trader (receives materials from intermediate traders or directly from both formal/informal recyclable collection systems | End of chain recycler (receives material from apex traders) |

FIGURE 4.2   Informal sector in recycling value chain in India. Credits: Swati Singh Sambyal, September 2021.

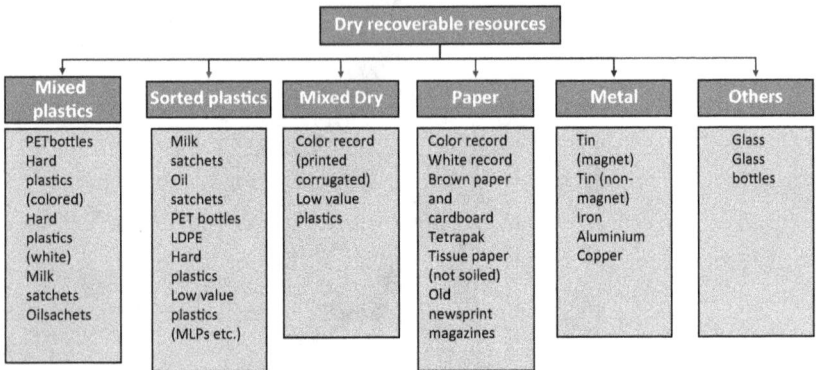

Mixed plastics	Sorted plastics	Mixed Dry	Paper	Metal	Others
PETbottles Hard plastics (colored) Hard plastics (white) Milk satchets Oilsachets	Milk satchets Oil satchets PET bottles LDPE Hard plastics Low value plastics (MLPs etc.)	Color record (printed corrugated) Low value plastics	Color record White record Brown paper and cardboard Tetrapak Tissue paper (not soiled) Old newsprint magazines	Tin (magnet) Tin (non-magnet) Iron Aluminium Copper	Glass Glass bottles

FIGURE 4.3 Categorisation of materials picked up by the informal sector.

4 INR per kg (refer Table 4.1). Also, these rates vary from city to city and state to state, and they depend upon factors such as the price of virgin material, collection and transportation, the availability of labour, and Goods and Services Tax (GST) rates.

As per one report, households, itinerant waste buyers, and waste pickers in India jointly recovered 1.2–2.4 million tonnes of newspapers, 2.4–4.3 million tonnes of cardboard and mixed paper, 6.5–8.5 million tonnes of plastic, more than1.3 million tonnes of glass, more than 2.6 million tonnes of metal waste, and 4–6.2 million tonnes of other recyclable material per year (Nandy et al. 2015).

4.2.1 Impact of GST on the Recycling Sector

The Goods and Services Tax (GST) further affected the resource recovery sector, majorly the informal businesses dealing with dry recyclable wastes. Scrap material was not taxable prior to GST, except for e-waste and metals which entailed a 6% Value Added Tax. After GST, all scrap material have a tax rate of mostly 18%.

Every apex trader or end-of-chain recycler with sales above 2 million INR per year has to pay GST every 15 days. However, there is also a chain of small or itinerant buyers and intermediate traders that is connected to such traders. The intermediate, small, or itinerant dealers are not formalised and exempted from GST and the waste generators also do not have a GST number. Hence, the apex trader/end-of-chain recycler must not just pay higher tax rates on his own margin but also for the value for which he purchases the scrap from all the other value chain actors. The recyclers also do not want to increase the final selling price of recycled goods to make it competitive to goods made from virgin material (refer Table 4.2). It is to be noted that previously, recycled

TABLE 4.1   Selling Price of Recyclable Materials in Delhi

| | PRICE (INR/KG) | | |
| | SMALL OR ITINERANT BUYERS TAKE MATERIAL FROM HOUSEHOLDS/OTHER SOURCES | | INTERMEDIATE TRADERS BUY MATERIAL FROM SMALL OR ITINERANT BUYERS |
MATERIAL	SITE 1	SITE 2	
Plastic (mixed plastics)	2–4	2–5	7
Carrier bags (Panni/ thaili)	2	2	3
Cardboard	9	7–8	9
Plastic bottles (PET)	20	15	20
Plastic containers/ buckets	15	15–20	25–40
PVC	3	3	20
Glass	1.5	1.5	5
Paper	1.5–1.75	1.5–2	11
Shoes	1–2	2	5
Wood	2–3	2.5–3	6–7
Glass bottle (per bottle)	1–2	1–2	1.5
Tyre	2–2.5	2	4–5
White paper	7	6–7	9
Iron	16–17	15	18
File	8–9	9	12
Thermocol	70	65–70	82–85

Source: Swati Singh Sambyal, Based on interactions with a few small, itinerant buyers, intermediate traders between August and September 2021.

goods had a cost advantage when compared to goods made from virgin material because they had much lesser taxes on them; for example, recycled polyester staple fibre (PSF) had a 2% excise duty while virgin PSF had a 12.5% excise duty.

4.2.2 What Determines Plastic Scrap Prices?

Retail markets in cities decide the daily 'market rate' of plastic scrap wherein traders are specialised in trading in wholesale quantities of items made from

TABLE 4.2 Inputs from Waste Pickers and Small and Itinerant Buyers on Current Rates for Materials Post-GST

MATERIAL	SMALL OR ITINERANT BUYER PURCHASES FROM WASTE GENERATOR (INR/KG)		SMALL OR ITERANT BUYER SELLS TO INTERMEDIATE/ APEX TRADER (INR/KG)	
	PRE-GST	POST-GST	PRE-GST	POST-GST
Newspaper	12	11	13.5	12
Iron	20	15	22	16
PET	13	10	14.5	12
Glass	1/bottle	3 kg	1.5/bottle	3.5/kg

Source: Data from Neha Walani, 'Will India's recycling sector collapse under the new GST regime?' Down to Earth, 8 August 2017.

recycled plastic grains. Information about the market rates is communicated through different mediums, from commission agents to scrap traders based across India. Once the price slab is set, based on the market rate, appropriate prices for plastic scrap are deduced at each stage of the plastic value chain (from manual sorting of scrap to the final recycled granules).

Price of crude and seasonal variation are the two key factors that influence the price of plastic scrap. Plastic scrap prices go down when crude oil price is low. Also, in winters prices are usually high in comparison to summer season, while during monsoons, market operations are hugely impacted as all trading activities get reduced. The scrap may also become wet which changes its bulk mass (due to seepage of water inside the bags), and therefore, cannot be sold. (Swati Singh Sambyal. 2019.)

4.3 SUCCESSFUL INFORMAL SECTOR INTEGRATION MODELS IN INDIA

One way to upgrade the informal sector would be to integrate waste pickers into directly collecting waste at source, with a right over recyclables and a guarantee of regular access to waste. This can be done by amending the existing byelaws on solid waste and plastic waste management, at the municipality level. Another approach could be integrating waste pickers into a formal cooperative or a microenterprise which is self-reliant and self-sustainable.

Also, circular businesses can play an important role in social inclusion of the informal sector by focusing on skill-building and integration of these workers. Some integration models ongoing in India can be characterised as social enterprise models (SWaCH and Hasiru Dala), circular businesses empowering informal workers (Phool), and municipal efforts to strengthen and integrate the informal sector in cities (the case of Ambikapur, Chattisgarh, India).

4.3.1 SWaCH Model, Pune: India's First Self-Owned Cooperative of Waste Pickers

The city of Pune in Maharashtra, India, significantly advanced its solid waste management by entering into a public–private partnership with the organisation SWaCH (Solid Waste Collection and Handling or, officially, SWaCH Seva Sahakari Sanstha Maryadit, Pune). In 2008, Pune Municipal Corporation (PMC) signed a 5-year Memorandum of Understanding with SWaCH for collecting source-separated waste from households and commercial establishments, transferring the segregated waste to PMC-run vehicles at designated points, diverting all recyclables from source, and permitting waste pickers to charge a user fee directly from citizens (*see Figure 4.4: Intervention of SWaCH in Pune*). The agreement also authorised waste collectors to retrieve and sell recyclables from aggregated waste.

Pune generates about 1,600–2,000 tonnes of solid waste per day. SWaCH covers 70% of the geographical area of the city, including surrounding villages, providing door-to-door waste collection services (Figure 4.5). The SWaCH door-to-door collection partnership has saved PMC about 1.13 billion INR each year and has reduced carbon emissions significantly through reduced truck usage. In 2016, the agreement between PMC and SWaCH was renewed for another 5 years with further expansion of the work in 2019 (refer Table 4.3). In 2020, SWaCH diverted 80,000 tonnes of waste to recycling. SWaCH members collect monthly user fees ranging from 50 to 70 INR per household and 140 INR per commercial entity for waste collection services. The total estimated cost of primary collection to the city is one of the lowest in India, at about 5.6 INR per household per month in 2020.

In 2020, the Coop Scrap Stores were launched as a unique initiative as these stores are owned and operated by waste pickers. They are fully self-sustainable, giving waste pickers an annual bonus of 10%–14% per annum of the waste sold by them. Another initiative also supports the collection and recycling of multilayered packaging (MLP) and other Low-Value Flexible Plastics (LVP) — The SWaCH-ITC Limited initiative is an example of an EPR-based system designed to ensure the inclusion of waste pickers and informal waste

1996
ID Cards to Waste Pickers

2005
Pilot Integration of waste pickers (10,000 properties)

2007
establishment by PMC & KKPKP

2008-13
1900 waste-pickers covering 3.3 Lakh properties

2014-15
Service without contract

2016
Second 5-year contract cycle

2018
1 lakh new neighbourhood coverage

2019
8.4 lakh covered (70%), including 11 new villages

2020
95% operation in COVID-19 Lockdown

Expansion to 100% Zero Waste

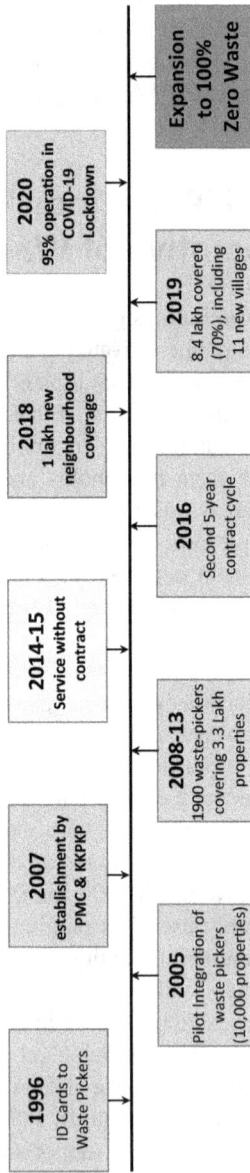

FIGURE 4.4 Intervention of SWaCH in Pune. Credits: Harshad Barde, SWaCH, 2021, September 2021.

TABLE 4.3   Current Status of Resource Management by SWaCH in Pune

Properties covered (all) (lakhs)	8.55
Slum properties covered (lakhs)	1.67
Total percentage coverage (%)	70
Total waste handled (TPD)	1,250–1,450
Diverted to recycling (TPD)	220
Segregation (%)	98
Waste pickers	3,650

Source: Data from SWaCH Pune, 2021.

FIGURE 4.5   Door-to-door collection of segregated waste by SwaCh workers. Picture Credits: SWaCH, 2021.

workers. Under this system, multilayered packaging (MLP) waste is directly procured daily from over 1,000 waste pickers through a fleet of mobile collection vehicles. The intervention was designed to minimise the need for storage and transport of MLP by waste pickers and incentivise resource recovery through a fair per-kilogram price of 4 INR. Usually, MLPs do not fetch any price or go with mixed waste at 2 INR per kg, hence having no economic viability. Currently, this system is being reshaped to integrate informal scrap dealers (Anantakrishnan. 2021).

To provide financial resilience to the public–private partnership, PMC provides an ongoing annual grant to SWaCH that covers management and training costs, awareness-generation programs, and welfare benefits for members of SWaCH. The grant does not cover the salaries of collectors. Through the partnership with SWaCH, Pune has offered sustainable and efficient daily waste collection services to residents while improving the livelihoods of waste collectors within the city.

Another innovative circular business model is The Protoprint Project (TPP), being implemented in Pune (Switch Asia. 2021). TPP adopts a collaborative step by step approach to reshape the informal recycling sector for downstream innovation of plastic waste. TPP sets up processes that support self-managed processing units consisting of waste-picker members. It uses affordable and standardised technological solutions to convert plastic waste into plastic flakes which are then sold to end users via negotiated agreements with industrial partners, assuring fair wages and buy-back assurance for the workers. This ensures each unit functions as a sustainable, replicable, and scalable business that can in due course be leveraged to integrate its informal workers into the formal economy; also providing benefits such as healthcare.

4.3.2 Ambikapur, Chattisgarh: Cities Integrating Informal Sector to Generate Value From Resources

The Ambikapur model is a low-cost, self-sustainable model set up in 2014 by the district administration with focus on three core areas – participation of women (informal workers), viability, and replicability.

The Swachh Ambikapur Mission City Level Federation (SAMCLF) started in 2015 as a unique project initiated to clean up the city. All the women in these self-help groups (SHGs) belong to the marginalised sections of society, including informal workers, and generate hundreds of green jobs. The SHGs are federated into a registered society called Swachh Ambikapur Mission Sahakari Samiti Maryadit. The Ambikapur Mission City (AMC) has an agreement with this registered society (Purohit. 2018) (Figure 4.6).

FIGURE 4.6 Sorting of recyclable waste at a tertiary centre in Ambikapur. Picture credit: Swati Singh Sambyal.

As of 2019, around 447 members from 34 SHGs involved in 18 area-level federations (ALFs) are responsible for the door-to-door collection of waste from over 27,000 households and more than 4,500 commercial entities and other establishments (Figure 4.7). Approximately 45–50 tonnes of waste per day are collected and processed daily. Collected waste is taken to a secondary segregation centre. The workers segregate the waste into 24 categories. Listed vendors purchase recyclable nonbiodegradable material at fair prices fixed by the AMC. Thus, an ALF on average earns INR 400,000–500,000 per month from the sale of biodegradable and nonbiodegradable waste.

In addition, tertiary segregation centres have been developed for further segregation of nonbiodegradable items. These centres are responsible for segregating nonbiodegradable waste into 169 categories and generally handle waste that exceeds the capacity of secondary segregation centres.

The incremental implementation and engagement of various stakeholders at every step has led to the strengthening of this model and a change in behaviour towards waste management in Ambikapur. In addition, SHG members earn roughly INR 6,000/month, creating livelihood opportunities for hundreds of underprivileged women informal workers.

Overall, the project saves the municipal corporation 52% of SWM costs per year, of which collection and transport-related costs alone account for over 41% of the savings. In addition, operational expenses at the dump site declined due the decrease in the amount of waste being disposed off there. Moreover, user charges of up to INR 1,600,000 per month were collected by the city, and the sale of nonbiodegradable wastes generated a monthly revenue of INR 400,000–600,000.

FIGURE 4.7   Collection of segregated waste from households at Ambikapur by SHG workers. Picture Credit: Swati Singh Sambyal.

4.3.3 Hasiru Dala: Social Business Enterprise

Hasiru Dala works with the informal sector and enables entrepreneurship by improving the working conditions of informal sector waste workers by ensuring continued access to livelihoods. Hasirudala began in 2010 by being an unregistered body but became a trust in 2013. Hasiru Dala works with the Bruhat Bengaluru Mahanagar Palike (BBMP), the municipal corporation of Bangalore, in 59 wards with 30,000 waste workers directly/indirectly enrolled into the model.

Once an agreement for a particular location has been crafted, the BBMP constructs a Dry Waste Collection Centre (DWCC) and hands it over to a group of waste workers from the organisation. The DWCCs were planned along the concept of a local recycling centre that would ensure maximum resource recovery and material channelisation. The DWCCs are charged with facilitating the collection/purchase of all recyclable waste from residents, sanitary workers, and waste pickers or scrap dealers. Installation of DWCCs also includes the integration of waste pickers and informal waste collectors in the operations of these centres and encourages Extended Producers' Responsibility (EPR) for packaging materials that are not being recycled presently.

The DWCCs are further linked to apex traders, and thus the waste pickers sell the waste to the linked traders as part of the organisation itself, as opposed to selling to third-party wholesalers and waste buyers. Hasiru Dala and its waste workers are able to divert over 1,050 tons of waste for recycling every day, leading to savings of Rs 8400 lakh (84 crores) annually for the municipality (Chandran and Shekar. 2018).

4.3.4 Phool: Circular Business Models

Founded in 2017, Phool.Co is an entrepreneurial venture based on circular economy that utilises floral waste and converts it into products (vermicompost, incense sticks, cones). Phool collects nearly 12 tonnes of flowers daily from 130 venues, employing women from low-income backgrounds. With deep-tech research, the start-up has developed 'Fleather,' leveraging its flower recycling technology.

Phool, while dealing with flower waste, is also focussing on providing manual workers (mostly women) with clean and secure jobs with benefits of a regular salary (over 7000 INR per month), secure bank accounts and access to health insurance and a clean working environment. As these flowers are used for worship, they are considered sacred and hence not sent to dumpsites/landfill. Phool upcycles flowers collected from rivers via local women into useful products such as incense sticks, soap and vegan leather articles. (UNFCCC. 2018).

4.4 SOCIAL INCLUSION (SI) OF INFORMAL SECTOR IN DESIGNING CIRCULAR ECONOMY MODELS IN INDIA

Based on the above examples and successful case studies on informal sector integration in CE transition, it is clear that we can reconsider SI in India with the focus on the informal sector from three perspectives: (1) strengthening social enterprises and encouraging informal entrepreneurs, (2) development of circular business models that integrate informal sector and (3) and most important, a focus by city municipalities on inclusive informal sector integration to strengthen their resource recovery systems to foster CE.

The following are some pertinent recommendations to strengthen CE transition in India:

4.4.1 CE Transition in Informal Sector through Social Innovation

The informal recycling sector in India is diverting 4.7 million tonnes of plastic per year from the public waste collection system, conserving natural capital, identifying material and resource value, and finding innovative solutions to several challenges large businesses are facing in secondary materials management (Mathias Schluep et. al. 2016). This sector can be an example of how transitioning towards a circular economy will help business, governments, and society achieve the UN SDGs. For example, 'moving away from resource intensive processes and maximising the use of existing material' not only contributes towards achieving SDG 12 (Responsible Consumption and Production) but also SDG 8 (Decent Work and Economic Growth), 9 (Industry, Innovation and Infrastructure), 11 (Sustainable Cities and Communities), and 13 (Climate Action).

4.4.2 Frugal Innovation

The innovations developed by the informal sector at grassroots level are mostly for self-use or for their community members. The absence of any profit-seeking intention from the developed innovation in the resource recovery chain and the lack of knowledge about instruments such as patents make these innovations diffuse widely in the local settings where they are developed (Sharma, 2021). The impetus should therefore be on how to measure these innovative activities and encourage them to transition to SI/CE with a focus on just transition.

4.4.3 Role of Informal Sector in Product Life Extension

The extension of product lifetime results in a reduction of waste and in resource saving. It is important to understand the importance of the informal sector and their contribution to the repair economy, an integral part of the circular system. Repair can contribute to material efficiencies in a circular economy, but it needs to be affordable and accessible for consumers. To explore the same, community repair shops or locally run repair cafés can be explored in cities, creating jobs in the informal sector. This will recognise informal workers not just as mere value chain sorters/recyclers but will also lead to their integration into the resource recovery system in cities, all the way from collection and segregation, reuse of secondary raw materials, repairs, and upcycling/recycling.

4.4.4 Integration of Informal Sector in Cities

It has become imperative to acknowledge and recognise the informal sector as the pillar of SI/CE transition in India. To do the same, the following measures need to be taken:

4.4.4.1 Identification and Mapping of Informal Sector Stakeholders in Cities

It is crucial for cities to identify (with support from local NGOs/CBOs/voluntary organisations) waste pickers and informal scrap shop owners/kabadiwalas in their operational area, list them, and create socioeconomic profiles, containing the name, age, gender, and quantity of waste collected by every single one of them. Waste inventory tools can be utilised by mapping the quantities of resources handled by them. Once these details are listed, they should be published in the Urban Local Bodies (ULBs) website. Also, a matrix (refer Table 4.4) to map the social inclusion of the informal sector can be developed to monitor progress. This data will not only enable effective decision making to integrate informal sector in cities but will also help to plan circular action plans for cities or social businesses.

TABLE 4.4   Social Inclusion Matrix for Informal Sector

KEY INDICATORS	INDICATORS
Social	Minimum wage
	Minimum working hours
	No child labour
	Social security (PF and health insurance)
	Job tenure
	Social identity
	Bank accounts
	Occupational safety health and working conditions
	Housing conditions
Environment	Segregation at source
	Resource recovery
	Waste leakages
	Waste mapping and channelisation
Economic	Revenue generation from resources
	Value addition through formalising and compliance
	Economic viability of project

4.4.4.2 Training and Capacity Building of Informal Stakeholders

Once identified and listed, ULBs must strengthen informal players in the city through capacity building so that the waste-picker community specifically can be trained and hired in the repair centres/material recovery/integrated processing facilities. Job retraining or skill-building programs on circular models, in combination with social support programs such as in healthcare and child education, can support adult career transitions and minimise periods of vulnerability (Centre for Science and Environment. 2021). Local NGOs and Corporate Social Responsibility (CSR) intervention by private enterprises can support such initiatives. The progress of these initiatives also needs to be regularly monitored/audited by the ULB or concerned authorities to mainstream informal workers in the formal resource recovery chain.

4.4.4.3 Guidelines on Inclusion of Informal Workers in the Circular Economy and Resource Management

The Solid Waste Management Rules 2016 focus on the integration of the informal sector; however, there is a need for implementable guidelines that showcase the inclusion of informal stakeholders in the resource value chain. The guidelines must incorporate provisions related to issuance of identity cards; access to recoverable waste for collection, segregation, and sorting; access to personal protective equipment to minimise occupational hazards; right to necessities like water, sanitation, and facilities for clean living; and health insurance.

4.4.4.4 *Ensuring Fair Pricing and Wages*

The informal sector market is highly dynamic and there are drastic fluctuations in prices, thus affecting stability in income and the livelihoods of informal workers. It is imperative to address and prioritise the issue of fair pricing and wages to stabilise informal income. To mitigate this, ULBs can collaborate with local organisations (NGOs or CBOs) to upgrade waste workers' income prospects by generating sanctions to ensure fair pricing from waste dealers, or by providing fair buy-back of recyclables at the recovery centres or by supporting waste co-operatives to set up direct contracts with waste dealers. This has been observed in the SWaCH and Hasiru Dala model.

4.4.4.5 *Informal Sector Cooperatives and SHGs*

Local governments and NGOs can also support waste pickers in forming organised cooperatives that provide a strong bargaining position with stakeholders. Micro- and small enterprises and cooperatives help waste pickers increase the purchase price of their collected waste by negotiating with intermediaries and allow waste pickers to gain social recognition. There may also be opportunities to access infrastructure to provide additional value to the recyclables, such as through bailing or cleaning the materials. A formal recognition, as observed in the case of Ambikapur, also provides for a stable income and livelihood for the informal workers including due appreciation of their work. Formal co-operatives (SHGs) also provide for a better standard of living and improved self-reliance and self-esteem over independent informal workers. With route optimisation and secure work conditions, organised workers are more productive and healthier.

4.4.4.6 *Occupational Health and Safety*

Considering most of the informal workers are not formalised or registered, their occupational safety and health is highly compromised. ULBs must ensure that personal protective equipment (PPE) such as helmets, gloves, rubber boots, safety goggles, masks, sanitiser, etc. are provided to these workers in addition to their registration. Under the purview of the enrolment of informal workers under a cooperative or a SHG, their health insurance must be guaranteed. The workers should be further trained for the regular usage of such PPE, and timely replacement of damaged PPE should be ensured by the ULB, NGOs, or the private CSR enterprise. In the wake of the pandemic, handwash stations must be installed in recovery centres.

4.5 INFORMALITY AND THE LEGAL REGULATORY FRAMEWORK

Most informal waste workers are neither in a relationship of employment with the households whose waste they collect nor are they in a relationship with the municipalities whose waste they sort and segregate. Cities must focus on incorporating/formalising waste pickers as workers working in the resource recovery industry (WEIGO, 2013). Additionally, providing waste pickers with identification cards can legitimise them as service providers for recycling operations in a city. Signing contracts with informal sector SHGs/co-operatives/enterprises gives their work a legal recognition and backing.

REFERENCES

Biplob Nandy, Gaurav Sharma, Saryu Garg et al. 2015. Recovery of consumer waste in India: A mass flow analysis for paper, plastic and glass and the contribution of households and the informal sector. *Elsevier*. Volume 101, August 2015, Page 167–181. https://www.sciencedirect.com/science/article/abs/pii/S0921344915300082

Gautam Sharma. 2021. Why India's huge informal sector needs its own innovation measurement metrics. https://science.thewire.in/the-sciences/why-indias-huge-informal-sector-needs-its-own-innovation-measurement-metrics/ (accessed October 18, 2021)

Govindan Raveendran and Joann Vanek. 2020. Informal workers in India: A statistical profile. Statistical Brief No 24. August 2020 WEIGO. https://www.wiego.org/sites/default/files/publications/file/WIEGO_Statistical_Brief_N24_India.pdf (accessed June 15, 2021)

Indian Plastic Industry Report. PlastIndia Foundation. 2019. https://plastindia.org/pdf/PSIR-final-for-print-04-03-2020.pdf (accessed July 05, 2021)

Mathias Schluep, Michael Gasser, Arthur Haarman, Andrea Brown, Brendan Edgerton and Elena Giotto. 2016. Informal approaches towards a circular economy: Learning from the plastics recycling sector in India. 2016. https://

sustainable-recycling.org/wp-content/uploads/2017/01/WBCSD_2016_
-InformalApproaches.pdf (accessed September 11, 2021)

Kamala Sankaran and Roopa Madhav. 2013. Legal and policy tools to meet informal workers' demands: Lessons from India. WIEGO. http://www.wiego.org/
sites/default/files/publications/files/Sankaran-Legal-Policy-Tools-Workers-India
-WIEGO-LB1.pdf (accessed June 05, 2021)

Lubna Anantakrishnan. 2021. Reflections on EPR-based systems targeting multi-lay-
ered plastics. https://globalrec.org/2021/03/29/reflections-on-epr-based-systems
-targeting-multi-layered-plastics/ (accessed August 17, 2021)

Makarand Purohit. 2018. Waste away, Ambikapur shows way. https://www.indiawater-
portal.org/articles/waste-away-ambikapur-shows-way (accessed June 17, 2021)

Marianna Marchesi and Chris Tweed. 2021. Social innovation for a circular economy
in social housing. Elsevier, Volume 71, August 2021, 102925. https://www.scien-
cedirect.com/science/article/pii/S2210670721002110

Pinky Chandran, Kabir Arora, Marwan Abubaker and Nalini Shekar. 2018. Valuing
urban waste: The need for comprehensive material recovery and recycling policy.   https://hasirudala.in/wp-content/uploads/2020/09/Valuing-Urban-Waste
-2019.pdf (accessed June 27, 2021)

Promoting socio-economic transformation by empowering informal waste pickers for
production of 3D printing filaments in Pune. *Switch Asia.* https://www.switch
-asia.eu/project/protoprint/ (accessed August 28, 2021)

Richa Singh. 2021. *Integration of Informal Sector in Solid Waste Management:
Strategies and Approaches.* New Delhi: Centre for Science and Environment

S. Kapur-Bakshi, M. Kaur, and S. Gautam. 2021. *Circular Economy for Plastics in
India: A Roadmap.* New Delhi: The Energy and Resources Institute. https://
www.teriin.org/sites/default/files/2021-03/Circular-Economy-Plastics-India
-Roadmap_0.pdf (accessed June 21, 2021)

Silpa Kaza, Lisa C. Yao, Perinaz Bhada-Tata, and Frank Van Woerden. 2018. What
a waste 2.0: A global snapshot of solid waste management to 2050. *Urban
Development.* Washington, DC: World Bank. https://openknowledge.worldbank
.org/handle/10986/30317 (accessed June 15, 2021)

Shravanthi Kanekal. 2019. Challenges in the Informal Waste Sector: Bangalore, India.
2019. https://penniur.upenn.edu/uploads/media/03_Kanekal.pdf (accessed on
July 2021)

Social Inclusion In WM. Saahas Zero Waste. 2019. https://saahaszerowaste.com/blogs
/social-inclusion-in-waste-management-what-it-means/ (accessed June 15, 2021)

Swati Singh Sambyal. 2019. *Recycling Plastics.* https://cdn.cseindia.org/attach-
ments/0.97245800_1570432310_factsheet3.pdf (accessed June 2021)

UNFCCC. Phool India. 2018. https://unfccc.int/climate-action/momentum-for-change
/women-for-results/phool (accessed July 18, 2021)

Private Sector Experiences in Circular Economy Business Innovation: A Case Study on Mahindra

5

Anirban Ghosh, Shreya Sankaranarayanan, Rachna Arora, and Pavithra Mohanraj

Contents

DOI: 10.1201/9781003201816-5

5.1 INTRODUCTION: THE ROLE OF THE PRIVATE SECTOR IN THE CIRCULAR ECONOMY TRANSITION

Private sector businesses make up one of the most critical groups of stakeholders in driving a system-wide transformation towards a circular economy. As manufacturers and suppliers of most products and services in today's economies, these businesses control a number of aspects that impact the circularity of products across the entire product lifecycle and can also influence consumer choice and adoption of circular models/products. For instance, businesses exercise great control over the following elements:

- the choice of materials, chemicals, and additives in products;
- the design of products, including their modularity, durability, recyclability, repairability, etc.;
- the production processes that they employ and their resource efficiency and waste/by-product generation and reuse/recycling;
- the business models through which they provide users access to their products/services (for example, sharing, rental models, etc.).

The private sector also has several other roles to play in driving the circular economy transformation, including as innovation engines and deployers of capital. Large enterprises especially have the potential to influence businesses that lie within their supply chains, and can enable and incentivise their vendors /suppliers to adopt circular practices. They also have a critical role to play as anchors in collaborations with start-ups, governments, multilateral agencies, and philanthropic organisations to achieve circular transformation goals through interventions in aspects that lie outside their direct control.

Businesses working collaboratively with other industry players within a value chain, or within sectors, are also critical to accelerating the development of voluntary standards, guidelines, and targets (such as the Jeans Redesign Project led by the Ellen MacArthur Foundation (Ellen MacArthur Foundation 2021) and the UK Plastics Pact led by Waste and Resources Action Programme (WRAP) (WRAP 2022), as well as harmonising policy approaches. The private sector, in partnership with the government, can ensure that the movement towards a circular economy model is large-scale and systemic, and not restricted to isolated pockets of the economy and society.

Recent years have seen a tremendous amount of intent from the private sector to transition to a circular economy, particularly from large enterprises. This is reflected in a number of shared visions and commitments set out by businesses. For instance, the World Business Council for Sustainable Development (WBCSD)'s Vision 2050, developed in consultation with over 40 leading global businesses, lays out a shared vision of a world in which 9 billion people can live well within planetary boundaries by 2050, and articulates transformation pathways that companies can take to achieve this vision. The nine transformation pathways encompass many critical circular economy strategies and business models (WBCSD 2022). Another example is the Joint Statement from the Ellen MacArthur Foundation, which has over 50 global leaders from different stakeholder groups as signatories, including CEOs of some of the world's largest companies, including Danone, Barclays Bank, H&M, The Coca-Cola Company, Unilever, etc. The Statement recognises the circular economy as an important solution to the question of 'how' to build back better after the devastating effects of the COVID-19 pandemic, and calls for leaders to 'step up, and not step back' from their CE commitments (Ellen MacArthur Foundation 2022).

Private sector momentum towards adopting CE approaches is easily understood in the context of the commitments that these businesses are making to achieving climate goals and net-zero targets and the increasing recognition that circular models are an important lever in the climate action toolkit. Current efforts around adoption of renewable energy and energy-efficiency measures can only address 55% of the greenhouse gas (GHG) reductions needed to meet the Paris Climate goals. The remaining 45% of emissions are associated with making products and need to be tackled through circular economy approaches (Ellen MacArthur Foundation 2021). Adoption of CE models are also vital for the achievement of several Sustainable Development Goals, in particular Goal 12 on Responsible Consumption and Production.

As companies increasingly adopt circular economy approaches, it will be crucial for them to set targets and measure the impact of their CE initiatives. The development of standard indicators and methodologies to measure business progress towards CE, such as WBCSD's Circular Transition Indicators (WBCSD 2022) and the Ellen MacArthur Foundations' Circulytics Framework (Ellen MacArthur Foundation 2022), is thus a critical step in enabling wider private sector adoption of CE.

5.1.1 Circular Economy Transition and the Indian Private Sector: An Overview

The private sector has powered India's development since economic liberalisation in 1991 and has a critical role to play in mitigating climate vulnerabilities and achieving equitable sustainable development.

According to one estimate, India could create as much as $218 billion in additional economic value by 2030 (and $624 billion by 2050) by adopting circular principles. Circular models can also enable a 23% reduction in GHG emissions and a 24% reduction in the use of virgin materials by 2030 (Ellen MacArthur Foundation 2016). Underlining the potential of the circular economy in India, Mr Amitabh Kant, CEO, NITI Aayog, highlighted that CE has the potential to generate nearly 15 million jobs in the next 5–7 years as well as create lakhs of new entrepreneurs (Business Today 2019).

Transitioning to CE and capturing this economic value, along with the numerous social and environmental benefits, will require significant behaviour change, technology development, innovation, infrastructure, finance, and investments, which will not be possible to achieve without the active engagement of the private sector.

In India, most of the private sector activity towards CE has been generated by the large corporates, including domestic companies and multinational companies (MNCs), as well as start-ups and innovators. Figure 5.1 highlights a few illustrative examples of how private sector players are driving the circular transition in the country through multiple roles:

A recent analysis by Accenture (Accenture 2021) indicates that while Indian businesses are highly aware of the value of incorporating CE models and their role in decarbonisation, only 27 of the top 100 companies by market capitalisation have disclosed CE-related targets, while only 3 companies have disclosed targets on absolute material consumption. Out of the five business models under CE as defined by Accenture, circular inputs and resource recovery models have seen much higher adoption by Indian businesses, while adoption of models around product life extension, product as a service, and sharing platforms were reported to be much lower.

In the next section, we examine case studies from the Mahindra group of companies to understand business innovation towards a circular economy – how businesses can begin their circular economy journey, derive value from CE, and integrate CE into overall sustainability commitments. We also examine the importance of corporate culture and employee engagement, as well as challenges and recommendations from the Mahindra experience on creating a pathway for sustainable transition.

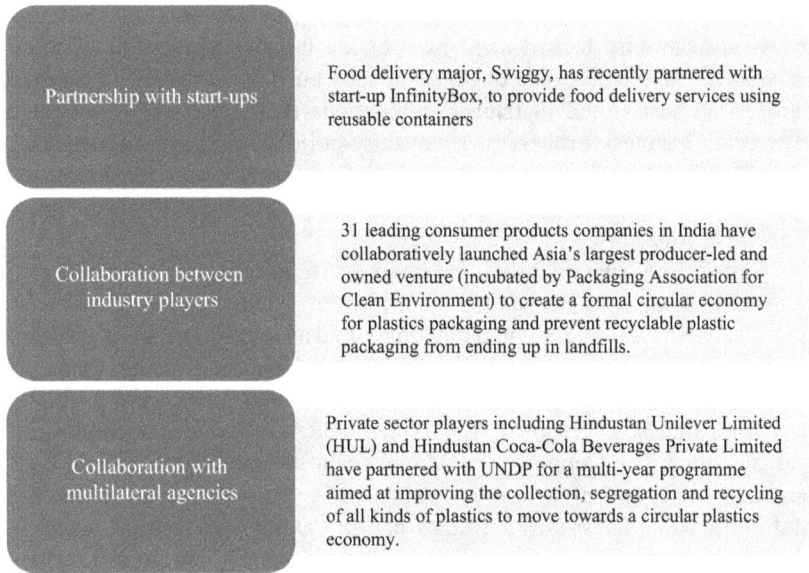

Partnership with start-ups	Food delivery major, Swiggy, has recently partnered with start-up InfinityBox, to provide food delivery services using reusable containers
Collaboration between industry players	31 leading consumer products companies in India have collaboratively launched Asia's largest producer-led and owned venture (incubated by Packaging Association for Clean Environment) to create a formal circular economy for plastics packaging and prevent recyclable plastic packaging from ending up in landfills.
Collaboration with multilateral agencies	Private sector players including Hindustan Unilever Limited (HUL) and Hindustan Coca-Cola Beverages Private Limited have partnered with UNDP for a multi-year programme aimed at improving the collection, segregation and recycling of all kinds of plastics to move towards a circular plastics economy.

FIGURE 5.1   Roles played by private sector businesses in driving CE in India [Figure created by the authors with information from Soni, Yatti. *Hindu Business Line*. 2021; *Hindu Business Line*. 2019; Ubuntoo. 2022; UNDP. 2022]

5.2 BUSINESS INNOVATION TOWARDS A CIRCULAR ECONOMY: A CASE STUDY ON MAHINDRA GROUP

The massive fire that broke out at the Deonar landfill in Mumbai, Maharashtra, in 2016 served as a wake-up call to many on the state of waste creation and management in the country. As people living in the vicinity of the Deonar landfill suffered the consequences of the fire for weeks, there was a clear realisation that society needed to pay attention to what was being sent to land-fills to prevent such catastrophes from happening again.

The incident at Deonar inspired Mahindra Group to find a way to address this issue through a logical solution. After internal discussions and clear delineation of aspirational initiatives versus practically implementable solutions, the Sustainability Team concluded that Mahindra businesses would ensure that they send no waste to landfill.

When the 'Ensure No Waste to Landfill' program was first initiated, teams at Mahindra lacked awareness of the broader concept of Circular Economy. Since employees engaged in core business processes were recognised as best suited to trigger innovations that are closely connected to regular business processes, a capacity-building programme targeting them was developed. As a result, many business associates were exposed to methods of analysis and insight development to enable the creation of innovative solutions.

Embedding sustainability in business required that employees at Mahindra understood the value of adopting environmental best practices to improve business processes and outcomes, and an overall culture of sustainability was built across Group companies, encompassing several elements of the circular economy framework. Environmental goals were promptly built into the Key Responsibility Areas (KRAs) of relevant employees, and Sustainability Councils and Management teams were set up in each business to review progress. A business excellence process called The Mahindra Way (TMW) used a Sustainability Dashboard to track progress in each company and ensure continuous improvement. Through sustained effort and regular engagement with senior management, businesses have recognised climate risks and started incorporating them into their risk registers. The internal audit team has also included sustainability processes and actions as part of its remit.

The Deonar fires occurred when the Mahindra Sustainability Framework was being discussed and developed within the organisation. The framework embraced the idea of simultaneously leveraging 'People, Planet and Profit' for balanced, sustainable business growth. The 'Ensure No Waste to Landfill' program was placed under the 'Planet' pillar of the framework and is one of the organisation's top ten Environment, Social, and Governance (ESG) commitments today.

Operations teams learnt that the goal to send no waste to landfill would break the take-make-dispose chain of the linear economy, and serve as a big step towards emission reduction through resource efficiency. Resource efficiency and circular economy activities, along with energy efficiency and renewable energy, were recognised as important elements of each Group company's journey to achieve Science Based Targets and become carbon neutral by 2040. Soon the business case for adopting CE approaches became very clear – not only could value be created from materials that were currently being discarded, but businesses could also prepare to deal with future scenarios involving scarcity of resources and extreme price volatility.

Ensuring zero waste to landfill is not limited only to the management of waste materials/by-products once they are generated, but begins with the manner in which a company designs its products, manufactures and distributes them, and enables productive use of materials in a product when it reaches the end-of-life stage. The closed loop approach provides opportunities to reduce, repair, reuse, repurpose, refurbish, remanufacture, and recycle, as appropriate, at each stage of a product's life cycle.

Transitioning from a linear to a circular economy requires a fundamental shift in the manner in which materials are regarded. Switching away from the perception of virgin materials as plentiful resources and instead regarding them as scarce commodities that are worth conserving brings a design sensibility that goes beyond cost and investments. Similarly, relabelling materials that remain unused during the manufacturing process as 'by-products' instead of 'waste' encourages the effort to create circular loops wherein the secondary materials remain in circulation rather than getting discarded.

In a sense, the circular economy helps us rediscover the values of conservation that are highly prevalent in societies where scarcity is the norm rather than the exception – societies which find new uses for old materials and discard things only when there is no option available for their reuse.

5.2.1 Embedding Circular Economy in the Mahindra Way

Mahindra was proactive in pioneering circular thinking in the private sector. Within a year of starting the 'Ensure no Waste to Landfill' program, its factory at Igatpuri, Maharashtra, became the first location in the country to be certified 'Zero Waste to Landfill.' In a short span of 4 years, 24 other Mahindra facilities emulated the Igatpuri model, including the Mahindra World City (MWC) at Chennai, which houses around 60 industrial facilities spread over 1,500 hectares.

The journey to 'Ensure no Waste to Landfill' generated many challenges and led to the development of customised solutions. The major steps undertaken were to assess the waste ending up at landfills, auditing downstream vendors with their processes, and assessing the potential to utilise waste products as secondary raw materials. Diverting non-hazardous waste from landfill proved to be easier than diverting hazardous waste. The Pollution Control Boards specify ways in which hazardous waste needs to be handled. While these regulations are well-designed for an ecosystem that is trying to prevent pollution, they are, at times, a bit restrictive for an ecosystem that is trying to enable a circular economy. One, at times, a bit of the most critical questions that emerged from circular thinking was focused on how recycled material can be incorporated in appropriate quantities into products, while ensuring that the finished product has greater recyclability.

Circular economy actions directed at managing waste are relatively easy to initiate as direct costs associated with regular ways of waste disposal can be reduced or eliminated using alternative methods for reuse as secondary materials. But the real value addition comes from reusing or repurposing the waste materials/by-products. The following sections discuss a few ways in which businesses in the Mahindra Group are transitioning towards a circular economy.

5.2.1.1 Value from Steel Scrap

Mahindra Accelo operates India's largest network of independent steel service centres and cuts large sheets of steel into shapes that are suitable for manufacturing processes in the automotive, power, and electrical stampings industries. Previously, it would sell the unused steel sheets as scrap, but by adopting CE thinking, the business identified a new revenue stream. It now manufactures cores for small electrical transformers from the unused steel pieces. The business is now a market leader in the power transformer segment with a market share of more than 40% in 2019.

5.2.1.2 Value from Waste Mud Generated in the Camshaft Grinding Process

The automobile business implemented its first Zero Waste to Landfill project at its engine manufacturing facility at Igatpuri, Maharashtra. To be certified as Zero Waste to Landfill, the factory would have to divert more than 99% of its waste from landfills. The waste mud produced during the grinding operation of manufacturing camshafts (a component in automobile engines) was proving to be a significant bottleneck in the process, considering its high volumes of generation (50 tonnes/year), its toxicity and categorisation as hazardous waste, and the additional costs for disposal procedures.

Several trials undertaken on utilising waste mud for road construction, paver blocks, etc., were unsuccessful, since the desired strength, toxicity, and quality parameters were challenging to achieve. Partnering with an external agency to recover metal, oil, and carbon black (an expensive raw material) from the waste mud proved to be a breakthrough.

5.2.1.3 Value from Food Waste: The Mahindra World City (MWC) Experiment

Mahindra Lifespace Developers Limited (MLDL) is a climate-conscious real estate business that builds green buildings and has embraced Science Based Targets for emission reduction. Apart from constructing buildings, it is the parent business for two integrated cities at Chennai and Jaipur, housing over 160 companies and thousands of residents. Mahindra World City, Chennai, is the first integrated city in India to achieve 100% diversion of waste from landfill and to be certified as Zero Waste to Landfill in 2016.

In 2014, MWC Chennai formulated a plan to divert food waste from landfills to reduce pollution and eliminate the need for the creation of a landfill. In partnership with engineers at the Mahindra Research Valley, the team at MWC Chennai set up a bio-CNG (Compressed Natural Gas) plant. The total waste diverted from landfill is ~135 tonnes per month, with approximately 115 tonnes of avoided CO_2 emissions, and 40 tonnes of compost and 3,000 kg of biogas generated every month.

5.2.1.4 Value from Food Waste: A New Green Business

Mahindra's experience with diverting food waste from landfill at MWC Chennai led to the formation of a new business, Mahindra Waste to Energy Solutions Limited (MWTESL), which until 2022 was a wholly owned subsidiary of Mahindra and Mahindra Limited. The MWTESL business converts municipal wet waste to bio-CNG and organic manure.

As of 2021, the business operated eight mid-sized plants, converting 20–40 tonnes of wet municipal waste to biogas at each plant and producing organic manure (rich in nitrogen, phosphorus, and potassium) using anaerobic digestion technology. The biogas generated at the plants is purified to become bio-CNG with greater than 90% methane content, and is used as auto fuel or for cooking applications. The digestate is converted into organic manure by adding biomass. The technology has been customised to use multiple bio-waste materials available in and around urban areas. The multifeedstock-capable biogas digesters can produce 2000–5000 m³ of gas per day. This business enables just transition by creating local employment opportunities, generating low-carbon renewable energy, and increasing soil fertility and resource conservation.

In 2022, MWTESL was divested as part of the Mahindra Group's overall capital reallocation strategy.

5.2.1.5 Value from Scrapped Automobiles: A New Green Business

According to the Central Pollution Control Board (CPCB), India had an estimated 8,731,185 end-of-life vehicles (ELVs) in 2015, with the number expected to go up to 21,895,439 by 2025 – an increase of 250% (CPCB and GIZ 2015). An old vehicle near the end of its life is eight times more polluting than a new one, so scrapping of old vehicles cleans the air. As mentioned in the CPCB report, informal recycling of ELVs does not follow environmental or legal norms and often provides unsafe working conditions for workers. Moreover, the use of primitive technologies for dismantling and shredding leads to low levels of scrap recovery.

Mahindra Accelo and Metal Scrap Trade Corporation Limited (MSTC), a Government of India enterprise under the Ministry of Steel, set up Mahindra MSTC Recycling Pvt. Limited (MMRPL), an organised auto shredding and recycling unit, to recycle ELVs in a clean and environment-friendly way. 'CERO,' the brand name for this business, stands for 'Zero' – zero pollution, zero wastage, and zero import of metal scrap (8 million tonnes of steel scrap is imported currently). At CERO's facilities, ELVs are depolluted, dismantled, and scrapped using world-class machinery and ensuring zero spillage of hazardous waste. Recovered steel, which is 70% of the car body, is converted to secondary steel, which would otherwise be imported.

5.2.2 Challenges and Learnings in the Transition

A circular economy is an entirely new ecosystem as it breaks conventional thinking, processes, and systems. It therefore creates risks for some existing businesses while providing new opportunities for many others. The first challenge is to establish the possibility of a better future using circular economy approaches through extensive consultation with all affected stakeholders and calibrated transition. Incumbent businesses in the waste ecosystem feel the threat of losing their revenue, and incorporating them productively in the new ecosystem is an important success factor.

A major challenge in the circular economy is the evolution of thinking, policies, and institutions from pollution prevention to value creation. The transition requires innovation through experimentation, something that is less important in a pollution prevention ecosystem. Initial failures, which are inevitable when running experiments, could make administrators wary of the transition, but it is critical to stay the course. As circular economy builds on the existing knowledge, skills, and competencies of industry and the regulators, and new technology and models emerge, partnerships between these stakeholders will make the transition happen faster.

One learning from recycling businesses is that the supply side is the biggest bottleneck – for CERO, getting a steady flow of vehicles to scrap and recycle is a challenge. This is similar to the experiences of the food-waste-to-energy business and is something plastic recyclers discuss regularly as well. In any recycling business, the reverse logistics ecosystem to make raw material available for recycling must be addressed while attempting to create value from waste.

5.3 CONCLUSIONS

Across the world, private sector activity towards the circular economy has been gaining momentum, particularly in the context of increasing business commitments to climate goals and sustainable development goals (SDGs). Every innovation in circularity, from product and process to business model innovation, is an opportunity for the private sector to do what it does best – move capital and competence to effect change productively.

The Mahindra experience in India shows that CE can be a win-win proposition for people, planet, and prosperity. There is immense scope for experimentation, innovation, and value creation in the transition, which are elements that businesses value. The Mahindra experience makes it clear that the private sector can play a key role in helping communities access the tangible and intangible benefits of a circular economy. It also highlights the need for constant experimentation and innovation in the CE journey, and the importance of building CE ideals into the culture of the organisation.

It is important to recognise that a lot of the momentum in the private sector is driven by large enterprises (domestic and international companies). Actively engaging India's micro, small, and medium enterprises (MSMEs), the backbone of Indian industry, and enabling them to keep up with and leverage opportunities in the circular economy transformation, will be a pivotal next step with respect to the private sector.

There is recognition from all quarters that no single organisation will be able to navigate the impending circular transformation on its own. Businesses need to work collaboratively with each other, regulators, policymakers, and other stakeholders to create a vibrant, synergistic ecosystem that can realise economic value, and social and environmental benefits, from the transformation to a circular economy. Collaborating for CE innovation is particularly vital (see Chapter 7 for a detailed discussion on this topic). Sustainability is a key ingredient for resilience, and CE business models can support the building of an inclusive, egalitarian society powered by the private sector.

REFERENCES

Accenture. 2021. "Beyond incrementalism: A pulse check on India's circular transition", Accenture, 2021, Accessed on January 8, 2022, https://www.ficcices.in/accenture.pdf

Business Today. 2019. "Circular economy could create 1.4 crore jobs in next 5–7 years: Niti Aayog CEO Amitabh Kant", Business Today, June 18, 2019, https://www.businesstoday.in/jobs/story/circular-economy-could-create-14-crore-jobs-in-next-5-7-years-niti-aayog-ceo-amitabh-kant-205306-2019-06-18

CPCB and GIZ. 2015. "Analysis of end of life vehicles (ELVs) sector in India", Central Pollution Control Board and Deutsche Gesellschaft für Internationale Zusammenarbeit (GIZ) GmbH, August, 2015, https://cpcb.nic.in/openpdffile.php?id=TGF0ZXN0RmlsZS9MYXRlc3RfMTE0X0FuYWx5c2lzT2ZFTFZzSW5JbmRpYTMfLnBkZg==

Ellen MacArthur Foundation. 2016. "Circular economy in India: Rethinking growth for long-term prosperity", Ellen MacArthur Foundation, December 5, 2016, https://emf.thirdlight.com/link/s0nl5u7tqsdr-ke3ly4/@/preview/1

Ellen MacArthur Foundation. 2021a. "Completing the picture: How the circular economy tackles climate change", Ellen MacArthur Foundation, Accessed on January 8, 2022, https://emf.thirdlight.com/link/w750u7vysuyl-5a5i6n/@/preview/1?o

Ellen MacArthur Foundation. 2021b. "The jeans redesign: Insights from the first two years", Ellen MacArthur Foundation, 2021, https://emf.thirdlight.com/link/m32pivncqxmc-gp46rn/@/preview/1?o

Ellen MacArthur Foundation. 2022a. "A solution to build back better: The circular economy", Ellen MacArthur Foundation, Accessed on January 8, 2022, https://emf.thirdlight.com/link/w68inpdhttkd-71ev4i/@/preview/1?o

Ellen MacArthur Foundation. 2022b. "Measure business circularity: Circulytics", Ellen MacArthur Foundation, Accessed on January 8, 2022, https://ellenmacarthurfoundation.org/resources/circulytics/overview

Hindu BusinessLine. 2019. "FMCG companies join hands to launch packaging waste management venture", *Hindu BusinessLine*, September 19, 2019, https://www.thehindubusinessline.com/economy/fmcg-companies-join-hands-to-launch-packaging-waste-management-venture/article29459986.ece

Soni, Yatti. Hindu BusinessLine. 2021. "InfinityBox partners with Swiggy to push re-usable food containers", Hindu BusinessLine, December 21, 2021, https://www.thehindubusinessline.com/companies/infinitybox-partners-with-swiggy-to-push-re-usable-food-containers/article37999938.ece

Ubuntoo. 2022. "Enabling circular economy pan India", Ubuntoo, Accessed on January 8, 2022, https://ubuntoo.com/solutions/karo-sambhav-1

UNDP. 2022. "Plastic waste management programme (2018–2024)", United Nations Development Programme, Accessed on January 8, 2022, https://www.in.undp .org/content/india/en/home/projects/plastic-waste-management.html

WBCSD. 2022a. "Circular transition indicators (CTI)", World Business Council for Sustainable Development, Accessed on January 8, 2022, https://www.wbcsd .org/Programs/Circular-Economy/Factor-10/Metrics-Measurement/Circular -transition-indicators

WBCSD. 2022b. "Vision 2050: Time to transform", World Business Council for Sustainable Development, Accessed on January 8, 2022, https://timetotransform .biz/

WRAP. 2022. "The UK plastics pact", The Waste and Resources Action Programme, Accessed on January 8, 2022, https://wrap.org.uk/taking-action/plastic -packaging/the-uk-plastics-pact#

Start-ups and Digitalisation for Circular Economy in India and Europe

6

Reva Prakash, Abhijit Banerjee, and Stefan Šipka

Contents

DOI: 10.1201/9781003201816-6

6.1 INTRODUCTION

Innovation will be a central driver for the transition from a linear to a circular economy (CE). This chapter showcases innovations from India and developments in the European Union (EU), especially on digitalisation. It identifies and analyses salient aspects for creating an enabling and empowering ecosystem, including through mainstreaming CE in public policy, for CE innovation.

Start-ups are important drivers for the development of an innovative and dynamic ecosystem as they develop creative and disruptive solutions for unaddressed gaps, and often support open innovation practices (Spender et al. 2017). The opportunities for start-ups and the challenges experienced by them also provide insights into the state of an ecosystem, and the changes needed for increased effectiveness and efficiency. For CE transition, collaboration and partnerships through innovative models will be needed.

This chapter is divided into two parts. The first part covers three illustrative case studies, but is not a comprehensive or exhaustive review, of innovative start-ups in India at different stages of maturity of their business model and scale. It captures innovations across different sectors and highlights certain commonalities and differences in their experiences and journey. The focus area of two of these case studies – Recykal and Banyan Nation – is on the end-of-life phase of products. While Recykal tries to utilise the power of digital innovation to create greater value for stakeholders in the waste management sector by connecting stakeholders and facilitating transactions, Banyan Nation focuses on closed-loop recycling and contributes towards the increased use of recycled plastics for packaging products. Kiabza's model is focused on 'thrifting' and contributes to product life extension in the textile sector. Instructively, all the three models rely on digital platforms and tools for their operations.

The second part focuses on digitalisation as an enabler for the CE and captures the experience from Europe through product life cycle stages. It analyses the challenges to digitalisation and the policy framework of action needed to support CE and digital innovation.

6.2 CASE STUDIES FROM INDIA

Methodology
In the selection of the case studies, cases were shortlisted from different life cycle stages and sectors. Case selection was done based on different life cycle stages, sector, maturity (with at least 2–3 years of operation since inception), and market uptake of solutions by target audiences. Thereafter, on the basis of willingness to be showcased and availability, primary semi-structured interviews with the founders/managers of the selected start-ups were conducted. A questionnaire was designed covering the journey of the start-ups from the ideation phase to the current status; the motivations of the founders; their growth story, capturing early, mid-term, and enduring/current challenges; historical juncture; ecosystem landscape and its response; business model; demand and supply-side challenges; responses of central stakeholders/actors; and approaches to dealing with core circularity issues like traceability and transparency, investment, and profitability.

6.2.1 Kiabza

Kiabza is an innovative start-up that has pioneered resale of branded pre-owned clothing in India. In India, while there is a culture of clothing being used among family members or of informal exchange by the waghris community, a commercially viable enterprise based on pre-owned clothing is a new concept.

6.2.1.1 Origins, Motivation, and Rationale

Kiabza was founded in 2018 by a team with a successful ancestral family business in the Indian textile sector. The founders realised that there was no formal second-hand clothing market in India, unlike in the 'West,' where 'thrift stores' featuring used clothing are ubiquitous. In India, there is a long-standing stigma associated with used clothing among the middle and upper classes, who are also the biggest generators of textile waste. Clothing reuse through donations to poor people or nongovernmental organisations (NGOs) exist, but this represents a small fraction of the potential market. Kiabza's founders were motivated by the untapped business opportunity in marketing used clothing to the relatively affluent and aspirational classes.

Inspired by the growth in the high-end used clothing market in the USA, the founders commissioned a detailed market survey for India. The survey revealed that the younger generation (18–30-year demographic) were open to pre-owned 'branded' clothing, provided that quality, authenticity, and hygiene standards were met. Based on the survey results, the founders launched Kiabza, using their own capital as investment. They hired a suitable team, rented a warehouse, and contracted a software vendor for developing the e-commerce platform.

6.2.1.2 Business Model and Operations

Kiabza's business model contributes to product reuse/product life extension. Through a digital platform, customers interested in selling their used clothing contact Kiabza. Kiabza establishes contact with the customer to verify branding, quality, etc., and schedules a pick up through their logistics partner, which happens within 1–2 days. Upon arrival at the Kiabza warehouse, a detailed quality check and inventorisation are undertaken, followed by cleaning, photography, and online listing. Their software generates a selling price based on the type of clothing, brand, quality, etc. A seller account is created online and the products remain on sale for 3–6 months. Upon sale of the product, the seller receives payment; instant payment at a much lower price is rare. Upon completion of the sale, the same logistics partner takes care of the delivery to the buyer. Kiabza covers delivery costs for both seller and buyer, which is factored into their product pricing. Products that remain unsold after 6 months are donated to charitable NGOs. Accurate barcoding is used for inventory control and tracking throughout the process by Kiabza and their logistics partners. Along with facilitating e-commerce of used products, Kiabza also takes care of the quality control process, including cleaning and checking, to guarantee high-quality branded products to its buyers.

Through their logistics partner, Kiabza reaches buyers and sellers pan India. Most of the sellers are from the Metros and Tier 1 cities, while buyers range from Metros to Tier 1, 2, and 3 cities. All categories of textile products have a viable market – from the lower end 'fast fashion' to 'high street,' 'premium,' and 'luxury.' Kiabza's sales pitch – 'branded fashion at less cost' – resonates effectively with their targeted 18–30-year-old clientele. Sustainability concerns do not seem to be explicit drivers.

Currently, Kiabza operates from one warehouse in Mumbai with 16 employees. Initially, the e-commerce platform development was outsourced, but now IT and marketing are done in-house, and only logistics remains outsourced. Within the past 2.5 years, Kiabza's operations have grown from an average of 50 orders per month to about 1,000 orders per month. To date, an estimated 20,000 pieces of clothing have been re-sold by Kiabza.

6.2.1.3 Key Factors for Success and Challenges

Experience in the textile sector provided a solid foundation to the start-up wherein initial capital was self-invested. The market research results convinced the founders of the potential viability of their enterprise, especially since marketing pre-owned clothing was an entirely new concept in India. Hiring an experienced, competent, and motivated execution team was crucial to turning the innovative idea into a viable model.

Kiabza's future financial sustainability is tied to attracting capital investments. Currently, raising venture capital remains difficult in India for new concepts. With a proof-of-concept, Kiabza has felt the interest from potential venture capital investors increasing. Accordingly, their first capital-raising effort was planned for early 2020, but the COVID-19 pandemic has pushed that to early 2022. Achieving profitability will take a few years.

Kiabza's founders designed their business strategy keeping in mind their prior experience and market survey results, and they developed targeted and differentiated marketing including exploring celebrity endorsements. According to Kiabza, buyers have three main concerns – quality, authenticity, and hygiene – and its 'curation experts' have played a key role in guaranteeing customer satisfaction.

For expansion, Kiabza plans to explore partnerships with fashion brands to sell unsold/surplus items and quality control (QC) rejects at marked-down prices through its web platform. Pilots are planned in the near future and, if this strategy succeeds, it has the potential to decrease textile waste. This resale-as-a-service (RAAS) model has the potential to tap into the client segment which is not comfortable with buying 'pre-owned' but open to considering 'outdated' items from the previous year. Kiabza has also recently added a 'refashioned' category in which a product is redesigned in collaboration with designers and contributes towards product life extension.

Kiabza has not benefited from any government subsidies/incentives support so far, but it feels the need for the same. Also, more innovators working with the RAAS model may further aid their development. While this will increase competition for Kiabza, it sees it as important for the growth of resale models.

The Kiabza business model contributes to circularity in the textile sector through product life extension, i.e., keeping clothes in use for as long as possible through different consumers.

6.2.2 Banyan Nation

Banyan Nation (BN) is a recycling start-up that aims to recycle recovered plastic and ensure its quality for its reuse in mainstream products and packaging. BN's complete value chain approach, innovations in the post-consumer plastics supply chain, and proprietary plastic recycling technology adapted to the Indian waste scenario have enabled FMCG (Fast-Moving Consumer Goods) majors like Unilever and Reckitt to use recycled plastics in over 200 million packaging bottles as of November 2021.

6.2.2.1 Origins, Motivation, and Rationale

BN was established in 2013 by two partners with international experience, who wanted to use technological solutions to provide a quality recycling service in the Indian waste sector. Though fragmented, the Indian waste sector does a fantastic job of collecting and channelling post-consumer waste streams with high after-market value through an elaborate network dominated by informal actors. The 'higher value' postconsumer waste streams for paper, glass, and metals were already well optimised, so BN decided to focus on plastics. After gaining a thorough understanding of the sector, BN realised that although India had one of the highest recovery rates of waste plastics in the world, no mainstream brands and manufacturers claimed recycled content. This was because plastic recycling in India is dominated by small, informal recyclers who use crude, low-cost processes resulting in serious environmental, health, and safety hazards. Further, recycled plastics are typically of poor quality and contaminated with hazardous substances like lead, mercury, phthalates, and other additives, making it almost impossible for mainstream brands to use recycled plastics in their products and packaging. Within plastics, PET (polyethylene terephthalate) recycling is relatively well-established in India, so BN decided to focus on HDPE (high-density polyethylene) and PP (polypropylene), where there is room for tremendous improvement.

6.2.2.2 Business Model and Operations

Headquartered in Hyderabad, BN employs nearly 200 people, and has a current capacity of 5,000–6,000 tonnes/year, with plans to double capacity by June 2022. It offers responsibly recycled plastic at consistent volumes and prices through a 100% traceable supply chain – making it possible for major FMCG and auto brands to make recycled plastics a part of their product value chains. BN has also patiently engaged with the informal sector and developed a relationship of trust with their suppliers.

The HDPE and PP waste collected from the informal sector suppliers is shipped to Hyderabad and processed at their recycling facility, using BN's proprietary technology. This technology is adapted to the Indian post-consumer waste scenario and deals with the specific challenges of packaging waste in India, viz. removing product contaminants such as oils and creams, strong adhesives, non-standardised labelling, surface printed inks, dirt, grime, etc. High-quality recycled HDPE and PP produced is sold in granule form to manufacturers for packaging and products for major brands. A digital platform is also a key component of BN's processes.

A combination of BN's proprietary data intelligence platform for mapping and tracking informal plastic waste collectors and aggregators, its hub-and-spokes collection mechanism, and its deep on-ground vendor management teams have enabled the company to develop reliable supply chains of waste plastic. Part of the digital platform enables identifying, geotagging, and collecting critical volume and pricing data on informal sector collectors, while another part of the platform captures the quality details at each stage through photos, and maintains all relevant paperwork related to each trade with an informal sector supplier. This enables BN to have control over feedstock quality and volumes, visibility, and traceability of the supply chain to the last mile.

6.2.2.3 Key Factors for Success and Challenges

The company has recruited a team with a deep understanding of Indian plastics waste streams, its informal sector, and the needs of customers with global brands and commitments. BN's vision is to ensure high quality standards so as to build customer confidence in recycled plastics. The company has strict quality controls at various stages in the process, invests heavily in testing, and can incorporate customer specifications.

BN's innovative multifunctional digital technology platform, coupled with its elaborate network development with waste sector stakeholders, helped the company to establish a reliable supply of highly curated incoming feedstock, which is critical for maintaining quality and profitability. Furthermore, economic and political factors over recent years have pushed global brands to adopt circular production practices. BN also has had a first-mover advantage in the Indian context, and collaboration with FMCG giants like Unilever and Reckitt at the early stage provided credibility to its solution. Enhanced partnerships with buyers of recycled plastic (product manufacturers) and expanding its supplier base will drive its future growth. The Extended Producer Responsibility (EPR) notification, MoEFCC (February 2022) also focusses on mandating the usage of recycled content under the EPR plans for producers, importers, and brand owners. This will provide a further push to BN's business model.

However, in the absence of standard specifications, onboarding of new clients takes 6–18 months as each brand needs to individually test and qualify BN's materials for use. Given its experience of working with some of the largest international brands, BN has been able to reduce this 'onboarding' cycle significantly. Development of application-specific standards and specifications for the recycled plastic industry, similar to those that exist for virgin plastics, will provide impetus to innovators in the ecosystem including BN.

Raising initial capital was a challenge as it was difficult to convince potential investors about the viability of premium recycled plastics in the Indian market, which is dominated by low-quality plastic recycling by SMEs (small and medium enterprises). Customer scepticism about the quality and reliability of recycled plastics remains widespread, although it has reduced over time as more brands publicly claim the use of recycled plastics in their products and packaging, and customers respond positively to such changes. The company raised initial investments from high-net-worth individuals, and later from investors with strategic fit in the polymers and materials space, who understand the circular plastics market.

Unlike in the EU, where paying a higher price for recycled feedstock is common due to strong CE mandates, Indian manufacturers are much more price sensitive, and expect recycled plastics to be priced lower than virgin plastics. Recyclers thus find it challenging to compete with virgin material prices. BN tried to address pricing challenges through long-term volume and pricing contracts that smother the volatility in virgin material pricing. While this addresses the needs of customers in the FMCG industry, other industries that are more price sensitive have yet to adopt the use of recycled plastics.

The initial success of Banyan Nation has paved the way for many other entrepreneurs, especially as policy indications of moving towards EPR for plastic waste emerge. The company sees high volumes, innovation, and investments that address the specific needs of its customers as the key to competing in an increasingly competitive market.

6.2.3 Recykal

Recykal describes itself as a 'waste-commerce or w-commerce' company with the aim of utilising digital technology for a more effective waste management system in India. It has created a digital platform for connecting all major stakeholders interested in w-commerce, for dry waste categories including plastics and e-waste, with a claim of bringing about greater traceability and transparency in the transactions.

6.2.3.1 Origins, Motivation, and Rationale

Recykal was conceived in 2016 by three entrepreneurs who wanted to use the power of digital innovation and disruption to create greater efficiencies, and then zeroed-in on the waste management space. Two of the founders had prior experience of running businesses, including initiating e-commerce start-ups. Their experience and learnings provided them with valuable insights for running a successful start-up. After its conception, the first two years were spent understanding the ecosystem of waste management in India. The study findings revealed that inefficiencies and loss of value occur due to disintegrated and complex supply chains with stakeholders operating in silos. Recykal raised early-stage capital of nearly 1 million US dollars through angel investing by the founders and other investors, with trust being placed in the founders due to their earlier ventures. For the first two years, Recykal undertook market research for three broad categories of waste streams, viz. paper, e-waste, and plastics, including consumer behaviour towards waste disposal. It helped them gain an understanding of the value chains through interviews with major stakeholders, including waste collectors, urban local bodies, manufacturers, and waste collection centres, for different waste streams and the prevalent disposal and recycling practices.

The team found that a well-designed waste-commerce platform had the potential to address inefficiencies and provide more information to all actors (generators, aggregators, and recyclers) by connecting them through a marketplace.

In 2017, Recykal was launched. At first, the platform tried to enable collection and recycling through an app called 'Uzed' which connected consumers with the broader ecosystem. Subsequently, the company focused on building a pan-India marketplace that closely integrates the informal sector, waste collection centres, and recyclers and co-processors onto one digital platform.

6.2.3.2 Business Model and Operations

Recykal's business model provides a digital platform service to its users for their commerce in the waste sector, i.e., it is primarily a business-to-business digital waste commerce platform. Currently, it employs 110 people and has operations in 30 States and Union Territories, with linkage to more than 50 Producer Responsibility Organisations (PROs), 100 plus recyclers working in partnership with over 80 urban local bodies, and over 100 brands. The Reyckal platform, according to its website, has channelised 100 billion tonnes of waste since 2019. Recykal works with brands such as Parle-Agro, Hindustan Unilever, Amul, Pidilite, Marico, Bisleri, Panasonic, Tata Consumer Products, Mars, Godrej, Perfetti, and Heritage, among others.

All its major offerings are digital services for waste management stakeholder groups. Its key products are the Recykal Marketplace, EPR Loop, Smart Centre Solution, and DRS Solution. Its central product is the marketplace which gets supported by its augmented offerings of EPR Loop.

The marketplace serves to connect aggregators with waste processors/recyclers, and the users are charged a platform fee/commission. The platform allows for the predictability of demand and supply. Recykal is building enhancements in the platform to provide better visibility on the quality of the waste disposed, logistics optimisation, and better pricing, ultimately leading to increased waste commerce digitally. The on-ground team connects sellers of waste including informal waste collectors, small and big aggregators,[1] and collection centres, and supports them with training on using the marketplace app for listing the supply. Recyclers are also on-boarded on the marketplace. This ensures that the technology product is adopted by both the buyers and sellers of waste. The buyer is charged a fee based on the transaction value and is incentivised to adopt the marketplace offering as it ensures a consistent supply of quality material, reduces transaction costs by identifying sellers across India, and thereby allows recyclers to operate at a higher scale. As a key value add-on for a marketplace business, Recykal provides logistical support and quality assurance through a 14-step quality check process.

As the waste-commerce takes place through its digital platform, there is potentially greater transparency of waste flows. The transactions take place under formal agreements, and both buyers and sellers have access to a wider ecosystem, which provides for better/fair prices. Recykal claims that in absence of such a platform, the trade happens through existing networks and word-of-mouth, and remains limited.

The EPR Loop seeks to connect corporates with varied EPR targets, in different states, with PROs that meet their specific needs. Currently, EPR fulfilment lacks traceability, transparency, and impact mapping – all of which the EPR Loop claims to solve.

As per Recykal, it has the potential to provide an impact assessment of brand owners' EPR fee contributions in the ecosystem in terms of waste collected; the livelihood of waste collectors, waste aggregators, and recyclers; and the enhancement of recycling capacity in general.

6.2.3.3 Key Factors for Success and Challenges

Angel investment and strong market research enabled Recykal to develop products that could increase efficiencies. Given the potential that their products offer to different stakeholders, Circulate Capital provided Reyckal with a second round of funding. Until a successful revenue model emerges in a few years, Recykal will need to continually identify sources of capital investments.

The introduction of the EPR framework through the E-waste Management Rules 2016 and Plastics Waste Management Rules 2016 has had an immense impact on the waste management landscape and created an increased need for transparency and traceability of waste flows. This led to a greater acceptance and demand for Recykal's services and products, but it still needs wider adoption. Reyckal also provides digitalisation support services to its different stakeholders, including the aggregators, buyers, and sellers, to enable access to its w-commerce platform.

6.3 DIGITALISATION AS AN ENABLER FOR THE CIRCULAR ECONOMY (WITH EXAMPLES FROM THE EU)

Our modern economy and society are undergoing digital transformation due to the increased use of data and digitally enabled solutions. The reason for a strong push towards digital transformation is the pertinent efficiency gains, including improved information transfer, new business models, and enhanced industrial/waste management processes. If steered correctly, digitalisation can provide a strong boost to innovation and transition to a CE. The following sections explain, through examples, how digitalisation can serve as an enabler for the circular economy.

6.3.1 Information Transfer across Value Chains

Enhanced access to better information and improved connectivity between relevant stakeholders across value chains could greatly aid the transition to CE. This includes tracking and tracing of valuable materials, products, and components as well as dangerous chemicals down the value chain, to enable safer and more efficient reuse and recycling. Better data access held by private entities and public authorities can help citizens make sustainable choices and enable innovative business models. Digital tools such as online databases, digital tags, blockchain, and apps can greatly facilitate information sharing for the benefit of the CE.

- Substances of Concern in articles as such or in complex objects (Products) (SCIP) is the European Chemical Agency's database for information on substances of very high concern in different products. It aims to ensure information availability throughout the lifecycle of a product, including to recyclers, to enhance waste management.
- Information for Recyclers (I4R) is an online platform that enables waste operators to access information – provided by the producers – about the preparation for reuse and treatment of e-waste. I4R has been designed and developed by APPLiA and DIGITALEUROPE, equipment manufacturing associations, and is maintained and hosted by the WEEE Forum.
- TagItSmart allows stakeholders across the value chain – producers, customers, and recyclers – to track items and provide additional information. This solution uses Internet of Things (IoT) enabled by functional quick response (QR) codes to trace, track, and monitor fast-moving consumer products throughout the supply chain and their lifecycles.
- Circularise uses blockchain to improve transparency and communication across value chains. Its Smart Questioning technology enables stakeholders to ask questions on products via secure communication. It allows for efficient data sharing while also addressing the industry's concerns over data protection.

6.3.2 Service-Based Business Models

Providing product-as-a-service can help us achieve more with fewer resources, namely by reducing material consumption and minimising waste. Apps and online platforms can facilitate service-based business models by connecting users with service providers.

- Whim is an app developed in Finland that has benefited from open public data. It provides information about multiple transportation modes (e.g. train, taxi, bicycle) and allows customers to pay for the travel via the app. Users can either opt for monthly subscriptions

or the pay-as-you-go method. This multi-modal and service-based business model can reduce dependence on cars.
- Clothing-as-a-service online platforms are growing in Europe and beyond. Tale Me is a Belgian rental service for maternity and children's clothes. The Dutch brand MUD Jeans rents and recycles denim clothing. Urban Outfitters has started the Nuuly rental service.

6.3.3 Circular Design of Products

The design phase determines up to 80% of the environmental impact of a product and is central to ensuring the reparability, durability, and reusability of materials and products; and designing for disassembly, upgrades, and/or recycling. Digital twins, 3D modelling, and digital sculpting provide interesting prospects for improving design processes. Designers can use artificial intelligence (AI) to improve design processes by playing with numerous materials and structures and testing and refining design suggestions. Integrating digital technologies into product design (e.g., digital tags, blockchain) could help support information transfer across value chains as explained in the previous section.

- The Horizon 2020 Accelerated Metallurgy project aims to identify new metal alloys and create new materials via AI to reduce alloy development time by easing navigation through the different variables needed to create the materials. This project demonstrates how data and digitally enabled solutions could also be applied to the development of more sustainable materials.
- The Citrine Platform uses AI to accelerate the material design process to develop new chemicals and materials for high-performance applications (for example, in the aerospace industries). AI can learn from previous experiments, thereby improving traditional 'trial and error' processes and reducing the duration of the research process.

6.3.4 Circular Practices in the Use Phase

The use phase is important for the CE due to purchase choices that consumers or businesses make and how a product is handled (e.g., if it is being repaired). Digital tools such as online platforms, Internet of Things (IoT), and digital product passports can support greater circularity in the use phase by enabling access to relevant information to users and repairers.

- iFixit is an open-source online platform for repairing electronics and machinery. It contains repair guides, Q&A forums, and user-generated updates on existing and prospective equipment.
- Scan4Chem is an app developed within the AskREACH project. Consumers can use the app to scan the barcode on a product and request information from the supplier about the presence of substances of very high concern in the product. Having access to such information can empower consumers to buy more circular products.
- Bosch uses Zotrax M200 3D printers to produce spare parts for the machines and equipment on its production line. Such practices lead to product life extension by increasing the availability of spare parts.
- ThyssenKrupp gathers elevator data and uses IoT to enable predictive maintenance. Similar tools could be used to enable predictive maintenance of electronics such as printers or washing machines.
- Bundles has an online platform that lends home appliances. The use of IoT and smart algorithms enables the supplier to monitor machine performance and identify possible problems to predictive maintenance.

6.3.5 Circular Management of Waste

Once a product reaches the end-of-life phase, it is crucial to ensure that it is collected, sorted out, and dismantled to recover components and recycle materials. Digital tools such as sensors, AI, and robotics can considerably enhance end-of-life management.

- R-Cycle has developed a digital product passport which tracks and traces plastic packaging. The initiative resulted from the concern that plastics are often not recycled due to a lack of information about their recyclability. The digital product passport allows different stakeholders to provide information about plastic items, which is stored on a common data platform and accessible via digital tags (e.g., QR codes).
- ConnectedBin uses sensors and IoT systems for smart waste management. Waste collectors can access data (e.g., on waste quantities) from all the bins, thereby optimising waste collection.
- The Finnish ZenRobotics uses AI-supported robots for fast and precise waste sorting, including bulkier waste (e.g., construction and demolition waste).

- AMP Robotics has developed an AI-powered robotics system for e-waste sorting in Japan. Robotic arms, supported by cameras, sort waste with 99% accuracy and at a speed of 80 items per minute (i.e., four times faster than manual sorting).
- SUEZ uses advanced waste characterisation with multisensor data to improve its waste sorting and recycling. Infrared technologies enhance waste sorting, while digital twin technology enables sorting machines to learn from its digital images of waste items. SUEZ is also experimenting with blockchain technology to ensure the real-time traceability of waste flows.

6.3.6 Challenges to Digitalising the Circular Economy

Despite the progress that has already been made in using data and digital solutions for enabling a CE, several barriers remain to be addressed.

- Instructively, digital solutions are not actively developed to benefit the CE. Policies and investments have been slow to encourage purpose-driven digitalisation, where the use of data and the development of digital solutions would be geared towards achieving a CE. Due to a lack of framework conditions, knowledge, and skills, many businesses are slow to transform their businesses and models to benefit from digitalisation. Without guidance, incentives, and framework conditions, the full potential of digitalisation for achieving sustainability will not be realised. Without steering, digitalisation runs the risk of promoting a linear 'take-make-dispose' model and overconsumption.
- Data sharing between the different stakeholders across value chains is hampered by a lack of standards, concerns over data protection (including personal data and intellectual property rights), and trust. When data is made accessible, it is not always provided in a user-friendly form, thus making it less actionable.
- The cost of digital solutions, such as AI, 3D printing, and robotics can slow down their uptake. For example, 3D-printed shoes can cost hundreds of dollars, and using robotics for waste management requires significantly higher investments than conventional approaches. Likewise, digitally enabled service-based models may be more expensive than traditional approaches based on product ownership.

- Stakeholders such as consumers do not always have the awareness, literacy, or skills to use digital solutions to purchase more sustainable products or maintain and dispose of them in alignment with circularity goals. While apps can help connect, inform, educate, and even empower stakeholders, navigating the app market can be overwhelming for many people.

6.4 POLICY FRAMEWORK FOR ACTION

A policy and financial framework that supports the development of a digital CE should address remaining barriers with the help of data and digital solutions. While CE and digital agendas are promoted globally, the European Union's (EU) current focus on the twin, green, and digital transition could provide interesting lessons also beyond its borders – in aligning the agendas.

6.4.1 Global Developments on Circular Economy and Digitalisation

The UN's Sustainable Development Goals encapsulate the global ambition in terms of sustainable consumption and production (SCP). The global circular economy agenda is supported by the International Resource Panel and the Global Alliance on Circular Economy and Resource Efficiency launched in 2021. Also worth recognising is the role of global initiatives such as the Basel Convention, which restricts the movement of hazardous waste. Developments like China's restrictions on waste imports and the EU's decision to ban the export of its plastic waste also have a global impact.

However, globally, CE initiatives are insufficiently linked with the global digital agenda, whereby the latter is evolving independently with possible impacts for SCP. Standards for electronic data interchange (EDI) are developing, like the EDIFACT (Electronic Data Interchange for Administration, Commerce and Transport) and EDIFICE (Global Network for B2B Integration in High Tech Industries) regulatory frameworks. GS1 is another global initiative on developing standards – notably, on digital tags (e.g. radio-frequency identification and barcodes) – to enable information sharing. There is also global cooperation on developing blockchain (INATBA), IoT (oneM2M) as well as standards for the ethical use of AI (the Institute of Electrical and Electronics Engineers' P7000 series).

6.4.2 The EU's Digital and Circular Economy Policies

The EU is taking several concrete steps to promote CE and to benefit from digitalisation in these efforts. It is planning to develop a common European data space and a governance framework to drive the deployment of digital tools for information transfer, such as digital product passports. The proposed regulation on AI would establish a framework for promoting the use of AI in achieving environmental sustainability goals. New product ecodesign rules, inter alia for electronics, textiles, and furniture, are expected to make products more durable, repairable, and recyclable while exploring possibilities for linking these products with digital solutions (e.g. QR codes, watermarks). Proposed batteries regulation has already envisaged the introduction of digital product passports for industrial and vehicle batteries. Moreover, as the twin green and digital transition is recognised as a high priority under the EU's budget for 2021–2027, this should also result in support for the digital CE. While the EU's work has only started, these developments already indicate how the requirements for products in a European market could evolve. Given global value chains, the effectiveness of solutions like digital product passports will depend on international cooperation.

Notably, digitalisation will be important for the transition to a CE. An enabling policy framework (at global, regional, and national levels) that brings greater alignment between digital and circular agendas is needed that can harness the power of digitalisation for CE and greater sustainability.

6.5 CONCLUSIONS

Key insights can be drawn from the experience of the three start-up case studies as well as the experiences and examples of digitalisation and CE in Europe for further policy reform.

6.5.1 Digital Technology

The innovative use of digital technology by all three Indian start-ups discussed has been key to their success. For Recykal, its digital platform is its 'product,' offered as a service to clients. For Kiabza and Banyan Nation, who deal with physical goods, their digital platforms and tools are central to their operations. The platforms seamlessly integrate web-based and hand-held devices and are collectively able to offer a wide range of services – from buying and selling to tracking and traceability, quality control, and geo-location, as well as advertising. In the case of Banyan Nation and Recykal, the platforms have been developed for sufficient ease of use so that they have been widely adopted by informal sector partners who are key actors in the waste management space in India. A high level of sophistication and expense was necessary to develop and fine-tune these digital platforms, but without this investment, it could have been challenging to navigate highly heterogeneous actors, waste streams, and locations in a competitive manner.

Notably, digitalisation is now accepted as an important enabler of CE in Europe. Digital innovations, crucial to achieving greater efficiency, transparency, and connectivity between different stakeholders, have acquired maturity in the EU and are gaining traction in India. Enabling a policy action framework, both globally and nationally, is imperative for creating linkages on digital architecture and CE.

6.5.2 Policy Landscape

Government policies help shape the market, and CE mainstreaming is no exception. The recent adoption of digital platforms on the EPR notification on plastics released by the Ministry of Environment, Forest, and Climate Change in February 2022 has contributed in enhancing the role of digital technology towards better compliance and accountability. Sensing that the extension of EPR into plastic waste and end-of-life vehicles (ELVs) is imminent, plastic product manufacturers and automotive companies are taking cognisance of CE practices. There is broad consensus that formal recognition of the Indian recycling industry will provide greater impetus. Additionally, incentives such as preferential GST (goods and services tax) rates on recycled/recovered products can make a big difference.

6.5.3 Investment and Financial Viability

Attracting investment is central to the success of start-ups. The experience of all three start-ups examined here demonstrates that raising initial investment capital for CE ventures is extremely challenging in India. This is because of the predominant impression that the 'waste' sector is dominated by the informal sector, and formal businesses find it difficult to thrive. The experience of start-ups, including those not documented here, demonstrate that creating successful niches using innovative business models is possible. With good performance and market uptake, attracting capital investment may also become easier. This underlines the need for government incentives such as viability gap funding for CE start-ups.

6.5.4 Market Understanding

All three Indian start-ups identified a market niche through detailed market surveys, even in the face of challenging Indian market conditions in the waste sector. Kiabza identified a young demographic open to buying branded pre-owned clothing, Banyan Nation found that certain non-PET grades of plastics were undervalued in recycling, and Recykal realised that connecting corporates to PROs and other actors in the waste chain for EPR compliance could be a profitable service.

6.5.5 Partnerships

Patient and painstaking development of relationships with potential clients and partners, from big multinational firms to small informal-sector operators, without compromising high standards and ultimate goals is also key. Through persistent engagement, these start-ups have been able to convince potential partners of their value proposition and bring them on board.

As the policy landscape develops for the circular economy, the potential market for innovative start-ups, powered by proactive measures to utilise digitalisation for greater effectiveness, is likely to expand significantly in India in future. More successes will help to convince potential investors and business partners. Thus, pioneering start-ups like the case studies discussed here have paved the way for further expansion of this emerging sector in India.

NOTE

1. Together these actors are often colloquially called 'kabadiwallas' in many parts of India.

REFERENCES

Spender, J.-C., Corvello, V., Grimaldi, M. and Rippa, P. 2017. Startups and open innovation: A review of the literature. *European Journal of Innovation Management*, Vol. 20 No. 1: 4–30. https://doi.org/10.1108/EJIM-12-2015-0131. Accessed on October 15, 2021.

Models of
Collaboration for
Circular Economy
Innovation

7

Narendra Prasad Kolary and
Pavithra Mohanraj

Contents

DOI: 10.1201/9781003201816-7

7.1 INTRODUCTION

Collaboration between organisations to create and accelerate innovation has the potential to unlock a much larger pool of ideas, resources, and skills than innovating strictly within organisational boundaries. In one study covering more than 100 large organisations, companies that collaborated regularly with their suppliers showed close to twice the growth, as well as lower operating costs and greater profitability compared to their industry peers (Gutierrez et al. 2020).

The systemic transformation required to move to a circular economy (CE) model will require a rethink of almost every aspect of products, services, and business models in our current economic system and signals a tremendous opportunity for innovation. It will also require a reconfiguration of value networks and the way they operate in our current linear ecosystems. However, most organisations are unlikely to possess the entire gamut of capabilities and competencies to innovate along all these aspects to establish a viable circular proposition. Thus, unprecedented collaboration between all actors in a value chain, including the policymakers, financiers, researchers, and civil society influencing that value chain, will form the linchpin of this systemic transformation.

Collaborations targeted at innovations in CE are even more important considering that many CE solutions are only starting to emerge and mature, while a vast number of solutions are yet to be created. Innovations require significant time and resources for the testing of new technologies, products, or business models, as well as establishing their commercial viability and scalability. In this context, it is prudent to look to collaboration as a critical lever to significantly accelerate the pace at which CE innovations are created, commercialised, and scaled up.

Globally, several interesting models of collaboration for CE innovation are beginning to emerge. Such collaborations can accelerate progress along different stages of innovation. For instance, they can help with the creation and testing of new technologies, products, materials, and processes; piloting of early-stage solutions with stakeholders like industry and governments; and scaling up of proven solutions. Collaborations are also key to removing regulatory barriers with respect to innovation and facilitating sharing of knowledge across multiple organisations with varied areas of expertise. A few such models of collaboration are illustrated in Figure 7.1.

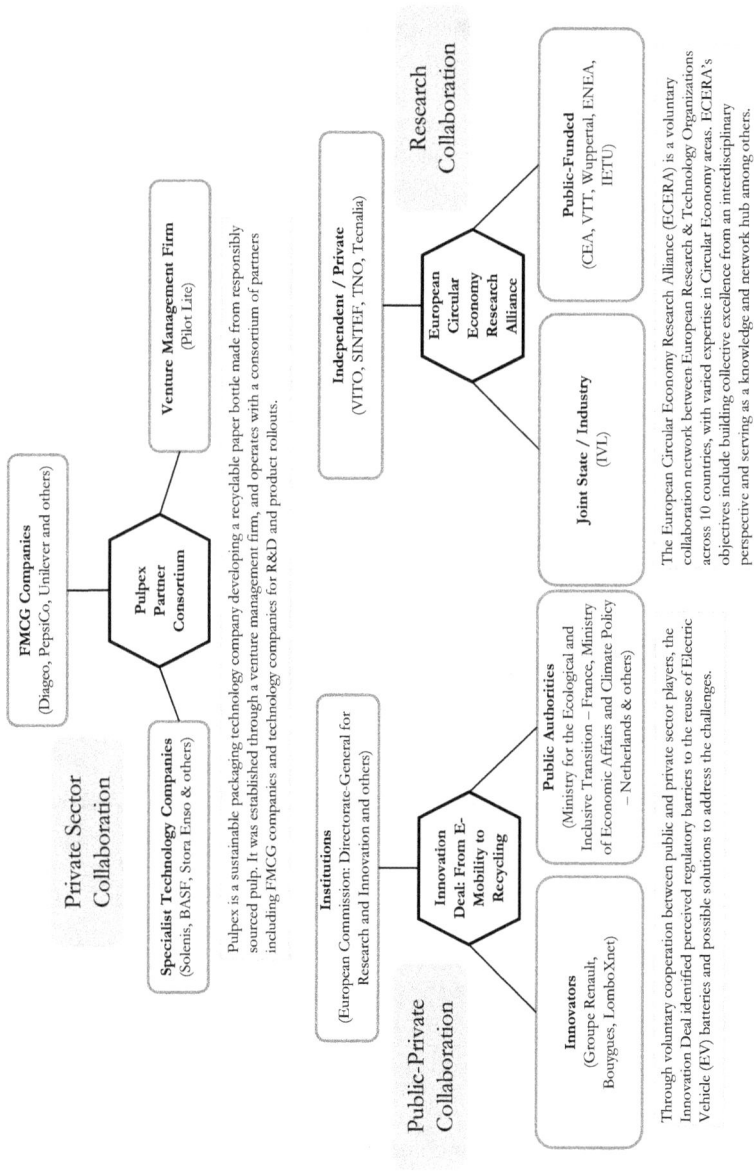

FIGURE 7.1 Models of collaboration for circular economy innovation. Figure created by the authors with information from Ellen MacArthur Foundation. 2021; Pulpex Limited. 2021; VITO. 2021.

With India being home to a rich entrepreneurial culture, many innovations for CE are stemming from the start-ups in the country. Collaborations that strategically bring together the respective strengths of start-ups and large enterprises have the potential to rapidly scale up promising innovations in a challenging Indian landscape. Start-ups display tremendous capability for disruptive innovation and are typically agile in their execution. Meanwhile, large enterprises bring to the table real-world expertise, extensive financial and non-financial resources, market presence and influence, and operational expertise (at a large scale).

There are several challenges that typically manifest when start-ups and larger industry players collaborate for innovation. These include:

- Nonalignment of expectations of stakeholders: Innovation pilots and projects are typically complex because of the varying expectations of each of the stakeholders. Setting expectations and boundaries for the collaboration is critical, but also challenging, given that start-ups and large enterprises are typically not on a level playing field.
- Difference in speed and flexibility of working: Innovators are typically smaller start-ups which are agile and fast-paced in their decision-making (usually founder-led or led by a small management team). This is in contrast to large enterprises which have multiple decision nodes and tend to be relatively less flexible in their decision-making.
- Capacity to scale: In collaborating with large enterprises or transitioning from small-scale pilots to larger-scale implementation, start-ups face challenges in rapidly scaling up their operations to meet the increased demand. Scaling a start-up requires funding, as well as teams with a different set of skills compared to the early growth phase, which takes significant time for them to build out.

7.2 CASE STUDIES FEATURING INDIAN INNOVATORS

Given the immense potential for disruption presented by start-ups, the case studies in this chapter have been chosen to showcase collaborative initiatives involving these young Indian companies. The case studies were also chosen to highlight examples of collaboration for CE innovation that differ widely in their sectors of application and stages of innovation. Cases were selected in such a manner so as to showcase one example each of multi-stakeholder collaboration and one-on-one partnerships, and to ensure coverage of initiatives that had a track record on execution. Case selection was also dependent on whether stakeholders were willing to share information for documentation. Since the chapter is not based on an exhaustive review of all such collaborations, the case studies chosen are only meant to be illustrative examples of collaborative models. The following case studies are discussed in detail:

1) *Early-stage innovation and multi-stakeholder pilot – fashion industry:* The first case study focuses on an early-stage innovation in the area of traceability for the fashion industry. The collaborative model involves an orchestrator bringing together a promising Indian innovator and key actors in the global viscose supply chain to pilot the innovation and test its real-world applicability.

2) *Growth-stage innovation and one-on-one partnership – plastic packaging:* The second case study focuses on a first-of-its-kind partnership between a growth-stage company, Lucro, and a well-established corporate, Dow. The partnership aims to leverage their respective expertise in commercialising flexible packaging with postconsumer recycled (PCR) content. This collaboration goes well beyond the typical vendor–supplier relationship between companies.

7.2.1 Case Study 1: Viscose Traceability Project

Collaboration between an Indian innovator and multiple value chain partners for piloting traceability solutions, managed by an orchestrator
This case study has been developed based on interviews with representatives from Fashion for Good and TextileGenesis.

7.2.1.1 Context

Textile value chains are typically complex, highly fragmented, and global in nature, with multiple actors involved in the product's lifecycle, from the processing of raw material and production to the retail phase and recycling/reuse of textiles. A lack of visibility of material flows across this value chain is a significant barrier to enabling circularity in the textile industry.

Transparency and traceability solutions, underpinned by blockchain technology, are an important cross-supply chain innovation and act as critical tools to track material flows for a product across its entire supply chain. Such solutions allow the verification of sustainable material sourcing and processing claims. They also capture information that is necessary to enable proper value capture of materials at the end of the product life cycle, and direct their further reuse and recycling.

Viscose has been growing steadily as a fibre of choice in the fashion industry for several reasons. These include increased consumer awareness around sustainability and the response from brands in seeking sustainable alternatives to traditionally dominant fibres such as cotton and polyester. However, about 50% of the 6.5 million tonnes of viscose currently used in global garment production comes from endangered forests (Canopy 2020). In order for viscose to gain acceptance as a truly sustainable fibre, it becomes imperative to ensure that the fibre is sourced from renewable sources and to establish mechanisms that allow verification of its origins.

7.2.1.2 The Collaboration

Fashion for Good (FFG) is a global, collaborative innovation platform (based in the Netherlands) that was established to tackle the sustainability and circularity challenges faced by the fashion industry. Fashion for Good works directly alongside the most promising innovators, bringing them together with market players (i.e. brands, retailers, and manufacturers) in order to make it easier for them to work together effectively, bridge the innovation gap, and bring these innovations to the mainstream.

From December 2020 to June 2021, Fashion for Good conducted a collaborative, global, multi-stakeholder pilot for testing a blockchain solution that enables traceability for viscose fibres. For the pilot, it brought together a consortium of key stakeholders from across the viscose supply chain. The digital solution was provided by Textile Genesis, a young start-up that has created a traceability platform custom-built for the apparel ecosystem. The platform allows brands to create article-level transparency from fibre to retail and verify the authenticity of sustainable fibres used in production.

As part of the pilot, the TextileGenesis traceability platform was used to track a total of 23,000 product units across 8 different garment styles for BESTSELLER and KERING (global fashion houses that manage multiple brands). The units were tracked across 25 suppliers in seven countries – Austria, Germany, Italy, Turkey, India, Bangladesh, and China. The pilot provided critical insights on how brands can employ such traceability solutions at scale and across different kinds of supply chains. Figure 7.2 provides a snapshot of the stakeholders involved in the pilot.

The pilot was able to successfully demonstrate the feasibility of applying a digital traceability solution globally across the viscose supply chain. The TextileGenesis platform proved to be able to capture real-world complexity in

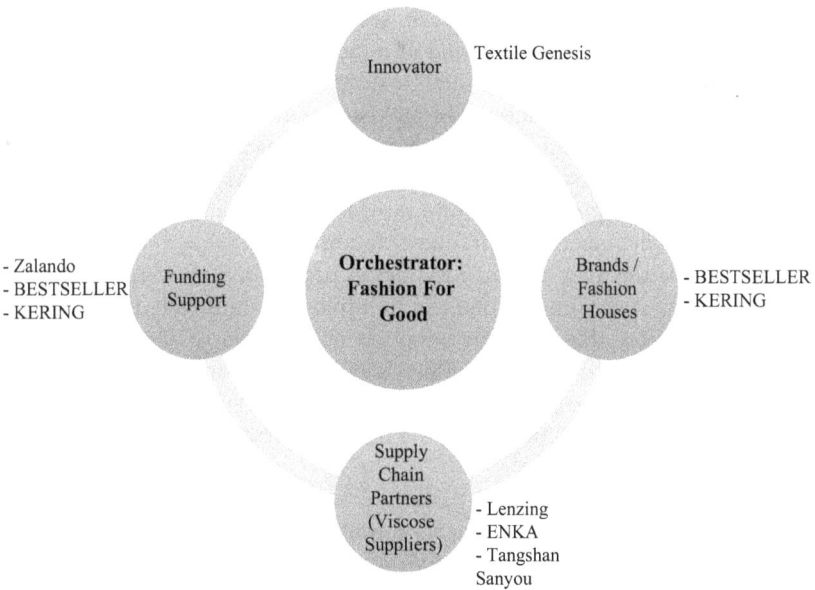

FIGURE 7.2   Key collaborators in the Viscose Traceability Project

supply chains, enabled by the Fibercoin tokenisation model. In this tokenisation model, every kilogram of physical fibre produced is represented on the platform as a Fibrecoin (a digital token) and is traced from the stage of fibre production until it reaches the brand. The platform also demonstrated interoperability with physical and digital systems – it was able to incorporate physical tracer technologies (such as those used by viscose suppliers Lenzing and ENKA) as well as interact with and incorporate data from other digital systems key to traceability (such as specific certificates from viscose producers). The onboarding of 25 different suppliers onto the platform within a 4–6-week timespan also helped to prove the scalability of the solution and its ease of use for stakeholders.

7.2.1.3 The Need for Collaboration: Multi-stakeholder Pilots

The transformation of the fashion industry to a more circular model requires a rethinking of all elements from design and production to retail models and end-of-use considerations. It will also require businesses to work with each other, and to also work closely with up-and-coming innovators. While we see some examples of corporates and innovators working together one-on-one (for example, as seen in Case Study 2), making the transformation happen at scale and at a rapid pace requires collaborative, multi-stakeholder pilots. Such pilots can move the needle on scaling and adoption of innovative solutions in lock-step with multiple actors in value chain. These are particularly relevant for early-stage innovations that need to be tested with industry players in a real-world scenario with commercial and operational implications before they achieve adequate credibility and mainstream acceptance as a circular solution.

Collaboration also becomes critical to propelling innovation forward, as many early-stage innovations can be almost prohibitively expensive in the initial stages, slowing their adoption at the level of individual businesses. The participation of multiple brands in a pilot with an innovator allows these brands to share the costs and risks of such experimentation, and is an important step in moving the industry forward as a whole towards adoption of these solutions.

While brands typically constitute the most visible parts of the supply chain and make public commitments towards sustainability and circularity, the actual implementation of technologies will have to take place at the level of the manufacturers and suppliers. Thus, the inclusion of manufacturers and suppliers in these pilots is vital for testing the effectiveness of an innovation and its real-world applications. Manufacturers and suppliers also possess considerable on-ground knowledge and experience that can provide critical feedback to improve an innovation or its applicability within specific constraints. Such pilots also pave the way for long-term relationships and commitments between brands and manufacturers/suppliers, which incentivises them to invest in more sustainable and innovative processes.

Also critical to these multi-stakeholder projects is a third-party entity that plays the role of 'orchestrator,' bridging the gap between industry and innovators. An 'orchestrator' like Fashion for Good is able to bring together brands and innovators, as well as intermediary players in the supply chain, to work collaboratively in a precompetitive space. Such collective investment from multiple actors in the value chain also allows for testing of solutions under a wide range of conditions and more open sharing of knowledge gained from the pilots.

7.2.1.4 Mechanics of the Collaboration

Selecting the innovation agenda for the pilot project
Fashion for Good typically works with its corporate partners to shape the innovation agenda for each pilot. This ensures buy-in and active participation from the corporate partners throughout the pilot. Beyond the pilot, it also allows

for implementation of promising pilot solutions within a brands' supply chain. Viscose is becoming an increasingly popular fibre, and supply chain transparency is becoming a critical issue for several brands working with these fibres. In this context, Fashion for Good and its collaborators conceptualised this pilot on traceability solutions for the viscose supply chain.

Selecting the innovator
The innovators in the pilots orchestrated by Fashion for Good are typically chosen from alumni of their Accelerator and Scaling programmes. Textile Genesis was part of the first batch of Fashion for Good's Asia Programme. It is a leading player amongst traceability solutions start-ups, having already demonstrated the technological feasibility of the platform and its ability to integrate chain-of-custody certification.

The TextileGenesis platform is based on blockchain technology and the GS1 traceability standard (a global standard used in healthcare and food industry) adapted to the textile industry. TextileGenesis' solution is based on an in-depth understanding of the intricacies of the textile industry, with nearly 18 months being spent on research and development (R&D) and grassroots engagement with the entire apparel supply chain for the development of the beta version of their product. Additionally, their custom-built solution for premium and sustainable textiles, as well as the pioneering Fibercoin technology, sets them apart from other traceability solutions in the market.

Selecting brands and suppliers for the pilot
Lenzing and ENKA are some of the largest viscose producers in the world and were already working with TextileGenesis prior to the pilot project. With supply chains that extend across multiple markets, they also provided the ideal test-bed to demonstrate the viability of the TextileGenesis platform.

Orchestrating and streamlining the collaboration
Bringing together collaborators in a precompetitive space is a challenging feat, and amongst other things, requires a highly qualified, credible, and trustworthy orchestrator. Despite being a young organisation itself (founded in 2017), Fashion for Good has quickly cemented its place as an important part of the innovation growth engine, and is well recognised for bringing together industry and innovators towards creating systemic transformation.

In multi-stakeholder pilots such as the Viscose Traceability Project, Fashion for Good adopts a number of good practices to ensure greater efficiency of collaboration and successful delivery of outcomes. Figure 7.3 highlights some of these good practices for orchestrating collaborative pilots.

Brand and manufacturer buy-in	•Key stakeholders must be certain of the positive impact of the project and need to be fully aligned on the project goals and outcomes.
Stakeholder roles	•The roles and activities of each stakeholder and the outcomes expected from each of them must be well-defined.
Well-articulated hypothesis	•The underlying assumptions leading to conceptualization of the project must be formulated into a hypothesis and have well-defined steps to prove it.
Innovation-centric project design	•The development of the project has to take into consideration the vision, capabilities and limitations of the technological innovations being tested. It is also important to bring innovators and industry players on to a level playing field.
Success criteria	•Success criteria have to be defined to enable proper evaluation of the project and determine next steps based on the success or shortcomings / learnings from the project.
Communications and reporting	•Regular check-ins with the stakeholders are needed to ensure that the project is on track and to continuously align activities.

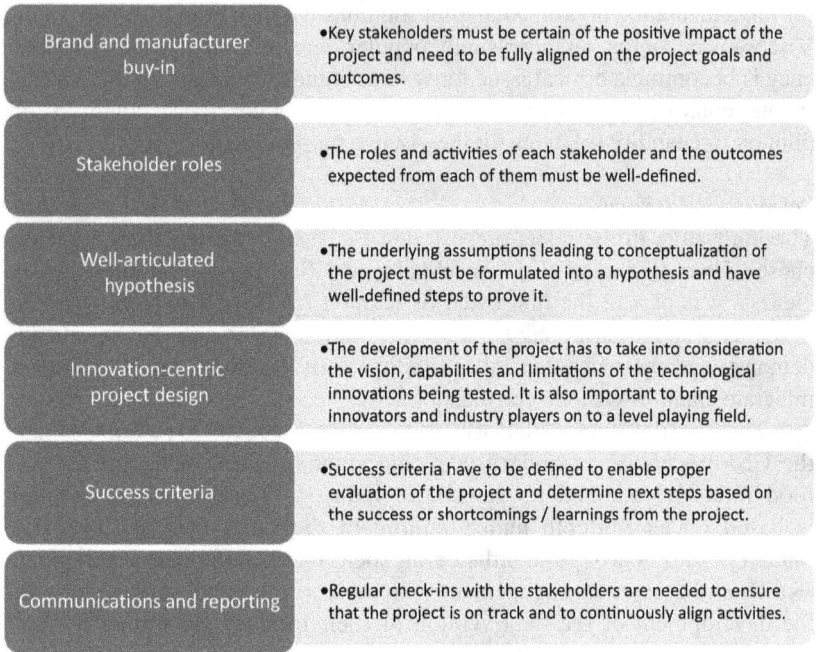

FIGURE 7.3   Good practices for orchestrating collaborative pilots

7.2.1.5 Next Steps

Following the successful demonstration of the capabilities of the TextileGenesis platform in traceability for the viscose supply chain, viscose producers Lenzing and Tangshan Sanyou will continue to participate in projects to scale these pilots. BESTSELLER will execute a second pilot in which they plan to trace one million product units.

Fashion for Good partners also plan to extend the implementation of the solution to other sustainable fibres, with six more fibre players having signed on for pilots tracing sustainable viscose, recycled polyester, and organic cotton.

7.2.2 Case Study 2: Dow–Lucro Partnership

Collaboration between an innovator and a corporate for incorporation of postconsumer recycled content in flexible plastic packaging
This case study has been developed through interviews with representatives from Dow and Lucro.

7.2.2.1 Context

Unlike other plastic streams such as polyethylene terephthalate (PET) that have well-established recycling networks in India, flexible plastics are difficult to recover and redirect into the recycling value chains due to the lack of recycling infrastructure and technology.

Over the last few years, increasing consumer awareness, existing and upcoming policies and regulations, and corporate commitments towards circularity have begun to shift the landscape with respect to postconsumer recycled (PCR) plastic in packaging. For instance, the latest Extended Producer Responsibility (EPR) notification (MoEFCC 2022) from the Ministry of Environment Forest and Climate Change (MoEFCC), Government of India, mandates the inclusion of recycled content in plastic packaging. However, flexible packaging remains an area that poses the most significant challenges and offers few workable solutions.

7.2.2.2 The Collaboration

Dow is a materials science leader committed to delivering innovative and sustainable solutions for customers in packaging, infrastructure, mobility, and consumer care. The company has set a target to enable one million metric tons of plastic to be collected, reused, or recycled by 2030 through its direct actions and partnerships. Over the last few years, Dow has been working across the value chain as a collaborative partner that cocreates circular economy solutions, including the design of recyclable packaging and PCR products.

Lucro (Lucro Plastecycle Private Limited) is an Indian company that has developed a proprietary Plast-E-Cycle™ process to convert flexible plastic waste into high-value recycled plastic granules. These are used in several packaging products including plastic carrier bags, automotive covers, shrink wraps, protective plastic sheeting, etc. Lucro manages the end-to-end process from collection and segregation of the waste to processing of granules, product design, and manufacturing of products from the recycled granules.

In early 2021, Dow and Lucro came together in a first-of-its-kind partnership in India to develop and launch polyethylene (PE) film solutions with PCR plastic content, with an initial focus on collation shrink films, a form of secondary packaging widely used by Fast Moving Consumer Goods (FMCG) companies. While Lucro has significant expertise and experience in the collection, processing, and manufacture of recycled resin from flexible plastic waste, Dow supports Lucro with its material science and application development expertise, for the development of a formulated film solution using Dow's virgin linear low-density PE and low-density PE resins with Lucro's PCR resins. This formulated film is manufactured and sold by Lucro. The work done through this partnership is expected to save 6000MT/annum of greenhouse gas emissions.

7.2.2.3 The Need for Collaboration: One-on-One Partnerships

The scale of the challenge involved in the widespread market adoption of PCR in flexible packaging is immense – ranging from the technological innovation required, to the creation and management of on-ground complex networks in the recovery and recycling processes, to the building of a reliable and viable business model. Additionally, the organisation(s) creating the solutions also need to have considerable credibility and influence in the market to persuade existing and potential customers to adopt these new solutions over tried and tested ones. Typically, in many emerging areas of innovation such as PCR in flexible packaging, a single organisation does not possess the technical, business, and operational skills and networks to develop, pilot, and commercialise solutions at a rapid pace.

The collaborative innovation approach seen in the partnership between Dow and Lucro is promising as a model that leverages different sets of expertise from both actors while significantly accelerating the pace of innovation and adoption, and its impact on the ecosystem. This approach offers several advantages to both the growth-stage innovator and established corporate:

Advantages for Dow

1) For Dow, this collaboration is aligned with Dow's sustainability targets of stopping the waste and closing the loop. By leveraging Lucro's existing on-ground network of plastic waste recovery and processing capacity, Dow brings its expertise in application development and polymer science, and the virgin resins necessary for blending with PCR to enable desired film performance.

2) Working with an integrated recycler like Lucro, which operates the entire value chain from the recovery of plastic waste to PCR resin production and recycled packaging products, simplifies the route from plastic waste to PCR films. The partnership creates synergy and efficiency for both companies in terms of goal alignment, strategy development, and execution.

Advantages for Lucro

1) Tapping into the materials science expertise at Dow has allowed Lucro to transform towards producing higher-value products compared to their initial offerings.

2) While Lucro has a robust in-house R&D team and facility, access to Dow's materials science expertise, along with Pack Studios

capabilities across three locations in Asia including India, has helped to shorten its product development cycle from an average of 3 years to as little as 3–4 months.

3) Facilitated introductions to Dow's customer networks have also significantly reduced the typically long sales cycle that innovators like Lucro traditionally face. The Dow partnership has also been instrumental in Lucro being able to gain access to the Indian market, while their previous sales have largely been to international markets.

4) The partnership with Dow has played a key role in boosting the market perception of Lucro as a trusted supplier of PCR-based packaging solutions.

7.2.2.4 Mechanics of Collaboration

Collaborations between a growth-stage company and a large corporate can typically prove to be challenging from an operational standpoint. Dow and Lucro follow a few good practices that have been important for navigating this collaboration:

1) *Alignment on priorities*: Start-ups typically tend to move fast but face challenges with the slower pace of decision-making at large organisations. In the Dow–Lucro collaboration, a clear alignment on goals and prioritisation of the circular economy agenda at Dow has been instrumental in keeping up the pace of the collaboration.

2) *Streamlining communications*: In addition to scheduled meetings for regular updates and communication, the teams at Dow and Lucro remain fully accessible to each other to support faster product and business development.

3) *Commitment of multiple teams and top leadership*: The global leadership at Dow remains invested in the collaboration. Multiple teams, from material science to business development, are also fully on-board with the requirements to keep the collaboration on track.

4) *Other value chain collaborations*: In addition to these good practices in managing their own collaboration, Dow has also been instrumental in facilitating other relationships for Lucro that are critical for a growth-stage company:

   • Dow connected Lucro with Circulate Capital, an investment management firm dedicated to fighting plastic pollution and advancing the circular economy in South and Southeast Asia. Circulate Capital invested in Lucro in early 2020. One of the founding investors of Circulate Capital Ocean Fund, Dow believes that by investing in local communities and entrepreneurs, it is possible to catalyse collective ambition to reduce waste in the environment and increase circularity by creating eco-friendly, sustainable goods.

- Lucro and Dow are working closely with many brand owners, such as FMCGs, to qualify and commercialise the PCR-based secondary packaging films. In 2021, Dow also facilitated a pilot project between Lucro and Marico, a leading Indian FMCG, to replace virgin shrink wraps on its Parachute[R] Oil product (hair oil product) with PCR. Following the successful pilot, Marico has recently commercialised Lucro's PCR-based collation shrink film with the use of Dow's booster and specialty resins for packaging.

7.2.2.5 Next Steps

Following the initial success of this collaborative approach, Lucro and Dow are working on scaling up the PCR solution:

1) After the successful pilot demonstration and commercialisation with Marico, Lucro and Dow are expecting multiple brands to commercialise the PCR-based film for secondary packaging.
2) Dow and Lucro also plan to expand their partnership to develop applications like shipping sacks, e-commerce bags, and non-food primary packaging.

7.3 CONCLUSIONS AND RECOMMENDATIONS

The case studies discussed above illustrate the need for collaborating for CE innovation and the impact that collaboration can have on accelerating the adoption of these solutions. The case studies also demonstrate a few critical success factors to ensure frictionless collaboration, as highlighted in Figure 7.4.

However, not all collaborations lead to successful outcomes. This makes it even more vital for stakeholders in the collaboration to clearly document their

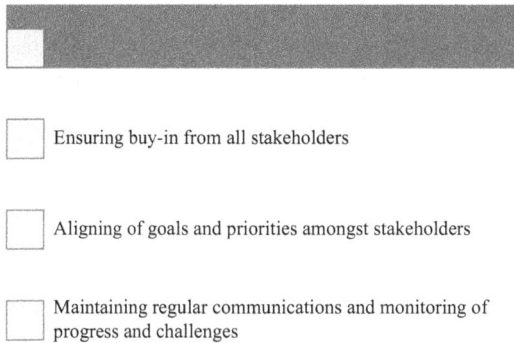

☐ Ensuring buy-in from all stakeholders

☐ Aligning of goals and priorities amongst stakeholders

☐ Maintaining regular communications and monitoring of progress and challenges

FIGURE 7.4 Critical Success Factors for Collaborative Initiatives

processes and learnings, in order to improve the design of subsequent initiatives and increase the likelihood of success.

Making transformative progress towards the circular economy in India will require many more such collaborations with innovators at different stages of growth, across different sectors and material streams, and in varied parts of the value chain. A few areas where such collaborations can have a profound impact in the Indian context are discussed here:

- A large number of innovative CE solutions are emerging in R&D centres such as technological universities, as well as amongst entrepreneurs. Structured mechanisms to validate these solutions and run collaborative pilots with industry players are key to improving the rate of conversion of these solutions from the R&D stage to commercialisation.
- Bringing industry players together in collaborative arrangements to sign long-term offtake agreements with promising innovators can simultaneously help to de-risk many challenges in adopting innovations while providing innovators with line-of-sight on the scale-up required.
- The policy landscape is often a critical enabler for the system-wide adoption of disruptive innovation. It is vital to bring together collaborative initiatives that engage industry, innovators, and policymakers to remove loopholes/barriers in existing policies and formulate new ones that can better enable circularity. A good example of this are the *Innovation Deals* (Ellen MacArthur Foundation 2021) formulated in the EU – these involve voluntary cooperation agreements between innovators and public-sector organisations to understand

and address nonfinancial barriers to innovation, including, for example, how existing regulatory systems are implemented and if they present impediments to novel and effective business solutions.

The following measures can help to increase collaborations to accelerate CE innovation:

7.3.1 Collaborating for Piloting Solutions

- Businesses are less willing to experiment with many early-stage solutions. Standardised mechanisms that can validate these solutions and improve their credibility in the perception of industry players need to be established.
- With collaborations being tricky to navigate, there is a need to set up more neutral third-party entities that can play the role of facilitators or orchestrators to bridge the gap between innovators and established companies. Multi-stakeholder consortiums that include innovators, industry players, academia, policymakers, international organisations, and capital providers and are managed by a neutral facilitator or orchestrator are critical to simultaneously draw out and address risks and challenges across all stakeholder groups and innovation stages.

7.3.2 Collaborating for Scaling Up Proven Solutions

To increase collaborations for scaling up solutions, there are measures that can be addressed at the level of large enterprises, at the level of start-ups, and at the ecosystem level.

7.3.2.1 Large Enterprises

- The most critical step in increasing the number and success of collaborative initiatives is to enable businesses to understand that thriving in the circular economy will require them to switch to a value-chain collaboration model. In addition to setting targets on circularity, businesses must be open to meeting these targets in a collaborative manner.
- Businesses need to examine how the models of value capture will change for their business in a circular economy. Based on this, they

need to delineate their strategic focus areas and strengths against other areas where they are better off working collaboratively with new innovators. Such an approach can enable businesses to create long-term win–win partnerships.

- Risk of failure of collaborative initiatives is a major deterrent for large enterprises that needs to be strategically addressed. For instance, to account for risks to the large enterprise that are related to the start-up scaling up, large enterprises would do well to ensure support for the overall growth of the start-up/innovator they choose to work with, including, for instance, providing connections to investors, improving operational processes, etc.

7.3.2.2 Innovators/start-ups

- There is a critical need to create a pipeline of robust and credible start-ups that are primed to collaborate with industry players and that can scale up rapidly according to market needs. Many start-ups face challenges in scaling up successfully due to a lack of capability of the teams, lack of expertise in the general market for emerging domains like CE, and operational and financial resource constraints.
- Start-ups have to be supported to navigate collaborations with larger companies including through support for legal, financial, intellectual property (IP) and other aspects that can become critical in a collaborative set-up. Ensuring fair share of value to the start-ups in their collaborations with large companies is vital. This could also ensure that more start-ups are open to collaboration, and do not see the large enterprises purely as a threat to their business.
- In engaging with corporates, start-ups have to account for risks from the failure of collaboration. They need to ensure that they are not completely dependent on the collaboration; and have enough internal capabilities and are sufficiently diversified to carry on their business even in case of the collaboration failing.

7.3.2.3 Ecosystem Level

- Structured mechanisms and platforms need to be created to provide all stakeholders with a snapshot of the possible opportunities and potential for collaboration, and to simplify the process of discovering potential partners.
- In-depth studies on the existing collaborations for CE innovation and showcasing the successes and challenges will be instrumental

in building a body of knowledge that can encourage more companies to pursue collaborations.

• Capacity building for both the start-ups and the large corporates to engineer and manage collaborations can help to prepare them for such initiatives and overcome the inertia to get started.

The circular economy is significantly different from our current linear economy and calls for greater collaboration and trust between stakeholders. Currently, very little is understood about models and mechanisms for collaborating for CE innovation, and it is an area that requires greater attention and incorporation into mainstream economic thinking.

REFERENCES

Canopy. 2020. "Survival – A pulp thriller: A plan for saving forests and climate", Canopy, 2020, https://canopyplanet.org/wp-content/uploads/2020/01/EXECsum-Next-Gen-Pathway.pdf.

Ellen MacArthur Foundation. 2021. "Unlocking opportunities through public-private collaboration: Using the EU's Innovation Deal mechanism", Ellen MacArthur Foundation, 2021, https://emf.thirdlight.com/link/gg4wzgfrzslv-img7o3/@/#id=2.

Gutierrez, Agustin, Ashish Kothari, Carolina Mazuera, Tobias Schoenherr. McKinsey & Company. 2020. "Taking supplier collaboration to the next level", McKinsey & Company, July 7, 2020, https://www.mckinsey.com/business-functions/operations/our-insights/taking-supplier-collaboration-to-the-next-level.

MoEFCC. 2022. "Guidelines on extended producer responsibility for plastic packaging: notification", Ministry of Environment, Forest and Climate Change, February 16, 2022. https://egazette.nic.in/WriteReadData/2022/233568.pdf.

Pulpex Limited. 2021. Accessed on December 6, 2021, https://www.pulpex.com/.

VITO. 2021. "ECERA: The European circular economy research alliance", Accessed on December 6, 2021, https://vito.be/en/news/ecera-european-circular-economy-research-alliance.

Emerging Trends in Circularity in India and the Impact of COVID-19

8

Pavithra Mohanraj,
Narendra Prasad Kolary, and
Mayuri Wijayasundara

Contents

DOI: 10.1201/9781003201816-8

In addition to the devastating health crisis that it caused, the COVID-19 pandemic has had wide-ranging effects on economies around the globe. It has led to supply-chain disruptions, business closures, and changed consumption patterns, while also adversely impacting livelihoods and societal development aspects.

The pandemic has also significantly impacted the nascent circular economy (CE) ecosystems that have been developing across the world. The fallouts of the pandemic have had both positive and negative influences on the uptake of the circular economy.

While many on-ground operations and new ventures in different stages of growth were impacted negatively, on the other hand, several large global businesses made renewed commitments to sustainability and the circular economy (Ellen MacArthur Foundation 2021). The economic recovery packages instituted by many countries to tackle the impacts of the pandemic also have circular economy strategies embedded in them – for instance, the Green Deal by the European Union.

In this chapter, we examine emerging trends and the impact of COVID-19 on two CE business models in India – subscription models for personal mobility and plastic waste recycling. Plastics (Accenture 2018) and mobility (Ellen MacArthur Foundation 2016) have both been well recognised as a priority material and sector, respectively, for the circular transformation in the country. Additionally, these areas experienced significant flux during the pandemic. The restrictions on movement of people and the impact on public transport forced users to realign their mobility choices. Waste generation increased for several kinds of plastics, while the recycling industry was negatively affected. Plastics recycling and mobility services have also been seeing a spate of entrepreneurial activity over the past few years. Finally, the chapter examines an example of distributed manufacturing in India. With COVID-19 exposing vulnerabilities in the way global value chains are currently set up, decentralisation has emerged as an important strategy for consideration in rebuilding our economies. [Research undertaken for this chapter is largely restricted to the period until December 2021, and may not reflect developments beyond this period.]

8.1 SUBSCRIPTION MODELS FOR PERSONAL MOBILITY

As India emerged from the stringent national lockdown imposed in the months of March–May 2020, safety and hygiene concerns surrounding COVID-19 were of paramount importance. As a result, public transportation as well as ride-hailing models fell out of favour with the public. The pandemic-associated disruptions also negatively affected the disposable income of many households in the country, and many consumers deferred their decisions to buy new vehicles (around 77% deferred in 2020 (Puthran 2021), and around 29% delayed or shelved their plans in 2021 (Choubey 2021)). Against this background, subscription models, such as those offered by start-ups like Zoomcar and Bounce, offered consumers safe and relatively affordable transport options. Subscription service providers reported a significant increase in demand in the period after the initial lockdowns were lifted, with some even reporting a 400% increase (Singh 2021, Lidhoo 2020).

A circular economy strongly encourages providing access over ownership for products to ensure reduced material consumption and more efficient material circulation. Subscription services for cars are an application of this model in the personal mobility sector. In a subscription service, consumers 'subscribe' to the car for a chosen period based on their usage needs at a fixed cost per week/month/year, rather than buying and owning the vehicle. In essence, the model allows users to pay only for the period for which they intend to use the car, giving them much greater flexibility than outright ownership. Additionally, the subscription model does not require customers to pay hefty down payments on the vehicles, and the onus of maintenance is also shifted from the customer to the provider of the subscription service. Since the ownership of the vehicle typically remains with the manufacturer or service provider in such models, it could create incentives for these stakeholders to improve resource efficiency and circularity across the product lifecycle from product design and use to disposal.

The period immediately after the national lockdown in 2020 saw a surge in the uptake of vehicle subscription models. This trend looks set to continue in a world where millennial preferences are shifting away from ownership and the uncertainties created by the pandemic continue to affect several aspects of people's lives. These include choices on where they live and how they commute, perceptions on the safety of public transport and ride-sharing/ride-hailing services, changes in purchasing power, and reluctance to undertake relatively long-term liabilities such as vehicle purchases. The subscription market globally is expected to grow at a Compounded Annual Growth Rate (CAGR) of 5–6 percent in 5 years (KPMG India 2020). This growth is expected to be particularly strong in India, where an overwhelming 69% of customers indicated that they are interested in a subscription service – more than in countries like China, the United States, and Japan (Deloitte 2021).

Start-ups and established Original Equipment Manufacturers (OEMs) alike see strong growth potential in the subscription model for the Indian market. Start-ups such as Zoomcar, Myles, and Revv offer subscription services with a choice of cars across multiple brands. In parallel, OEMs are also starting to offer subscription services across at least some of their models or enhancing existing offerings to cover a wider range of their car models and offer greater flexibility, shorter-duration plans, etc. These are offered by the OEM either on their own (Tata) (Tata Motors 2021), or in partnership with companies like Revv (Mahindra) (Revv 2021). A key signal pointing to an impending market shift may well be the model that companies like Zoomcar are starting to pursue, offering their technology platform for vehicle subscriptions (Zoomcar 2021) as a white label product for OEMs to use – for instance, in the partnership that they recently created with MG Motor (MG Motor India 2020). Passenger vehicle sales hit a 6-year low in the financial year 2020–2021 (Mohile 2021) amidst what was already a struggling auto industry before the pandemic. Against this background, subscription models are finding favour amongst established auto makers as new sources of revenue and as means to capture additional market segments.

On the consumer adoption side, access-over-ownership models typically face a number of challenges, including those of consumer awareness and behaviour change. A survey by CarDekho earlier this year highlighted that close to 70% of respondents (Pathania 2021) were unaware of subscription services for vehicles. In markets like India, with vehicle ownership traditionally being an aspiration and signal of upward mobility, significant behaviour change is also required for these models to gain acceptance. With the pandemic forcibly bringing about some of this behaviour change, there is great potential to leverage this momentum and look for ways to aggressively scale up these models.

As models like car subscription become more prevalent and accessible to consumers, there is a substantial risk of increase in the number of cars at the expense of public transport. This could in turn lead to an increase in pollution and traffic congestion, as has already been seen in the case of some ride-hailing models (Transport & Environment 2019). For subscription models to truly act as a solution towards a more circular economy, it is necessary to ensure that they do not lead to an increased use of resources as an unintended consequence, and that the entire system clearly focuses on reduced material consumption. There is also a need to constantly assess emerging solutions like subscription services and their impact on the overall circularity of the mobility system. It is necessary to have a well-planned mobility system in which pricing, incentives, and regulations are designed to prevent a 'rebound effect' – that is, to ensure that subscription models for personal mobility are not being chosen over public transportation (which is more efficient as a circular economy solution), thus unintentionally increasing the number of private vehicles.

To illustrate with an example, the current pricing points of subscription models in India would suggest that for most consumers, subscription is primarily an alternative to the purchase of new vehicles, rather than acting as a replacement for public transport (which is highly affordable in India). However, there is currently no clear data to substantiate this. Such data and assessments become even more essential in the face of the rapidly evolving landscape for subscription services that is likely to bring down price points and make these services appeal to a wider population. These include changes such as adoption of used cars and two-wheelers into subscription models. Simultaneously, there is a need to upgrade public transport infrastructure and quality of service to ensure its continued appeal and high levels of usage in the face of rapidly emerging alternatives.

8.2 PLASTICS RECYCLING VALUE CHAIN

Across the globe, concerns abound that the conditions created by the COVID-19 pandemic will reverse gains made in the decades-long battle against plastic waste. Around 8 million tons of pandemic-associated plastic waste has been generated from 193 countries as of August 2021 (Peng et al. 2021). Simultaneously, the conditions created by the pandemic have also adversely affected the plastics recycling industry, with recyclers reporting businesses shrinking by as much as 20% in Europe and 50% in parts of Asia (Brock 2020). These developments are shaping up against a background of increased investments in virgin plastic production, with major oil producers locking in investments to the tune of $400 billion in petrochemical plants (from which virgin plastics are produced) over the next 5 years (Bond et al. 2020).

Following the global trend, India has also seen a substantial increase in waste generation, including plastic wastes, in sectors such as healthcare and e-commerce. The increased plastic waste generation can be attributed to an increase in hospital wastes, a dramatic rise in the use of masks and other Personal Protective Equipment (PPE), as well as increased packaging material consumption from rising e-commerce sales, especially in urban centres (Aravind 2020, Accenture 2020). Bans on single-use plastics (SUPs) that had been instituted in several states across the country also saw their implementation become less stringent and take a back seat to the public health crisis (Vasanth 2021, Dharmadhikari 2020). Strides made towards raising public awareness on the detrimental impact of SUPs and reduction of their usage have also been adversely affected. Instead, SUPs have gained legitimacy amongst consumers as the better option to address the hygiene and safety concerns amidst the pandemic.

The plastic recycling industry in India, which had already been dealing with low prices for recycled plastics, saw complete devastation on account of COVID-19. With the oil price shock accompanying the pandemic outbreak, and its impact on the global economy, recyclers found themselves further unable to compete with the virgin plastic prices and were faced with reduced demand for their recycled resins. In the early months of the pandemic (April–May 2020), virgin plastic prices in India (and South-East Asia) decreased 30% year-on-year, paralleling rock-bottom crude oil prices. This in turn required recyclers to drop their prices by as much as 21% year-on-year compared to 2019 prices across different kinds of plastics (GA Circular 2020).

The Indian plastic recycling industry, made up mostly of the informal sector, was impeded by several other on-ground realities under the world's largest lockdown. In March 2020, a 21-day nationwide lockdown was imposed in India under the Disaster Management Act (2005), inhibiting the movement of ~1.3 billion people (Venkata-Subramani 2020). In addition to demand-side challenges, since recycling activities were not classified as 'essential services' by the Indian government, large parts of the recycling industry either closed down their operations or operated at very low capacities during the early phases of the lockdown. One estimate (GA Circular 2020) suggests that less than 10% of the recyclers remained operational during this period, with only 25%–50% utilisation of the total installed capacity post the lockdowns in June 2020. Of the 47 PET recyclers in India, 15 closed down during the pandemic (Suratman 2020). This near-total shutdown of the sector also jeopardised the livelihoods of roughly 2 million informal workers (Raveendran and Vanek 2020) as they found no buyers for the material that they segregated. The restrictions on people's movement for collection and sorting operations, concerns around the spread of the virus, and the migration of some of these informal-sector workers back to their hometowns also put further pressure on the recycling value chain and its ability to bounce back once the lockdowns eased.

It is generally well acknowledged that India's plastic recycling industry predominantly carries out downcycling – recycling of plastic in such a manner that the resultant plastic is of a lower quality/functional purpose than the material recycled. Even as the pandemic wreaked havoc on India's plastic recycling industry, several bright spots have begun to emerge, particularly in recycling of plastic packaging, which makes up 59% of plastic consumption in India (PlastIndia Foundation 2019). These developments could signal the beginning of a shift from low-value plastics recycling to higher value or closed-loop recycling in India. Some of these developments are discussed in the sections below.

8.2.1 Building Innovative and Resilient Business Models for Plastic Recycling

In October 2020, two leading Fast Moving Consumer Goods (FMCG) companies, Reckitt Benckiser and Hindustan Unilever (an Indian subsidiary of

Unilever), made announcements that signalled an important shift towards circularity in plastics packaging in India. Reckitt Benckiser unveiled a first-of-its-kind Dettol Handwash bottle made of 100% recycled plastic with resin procured from Banyan Nation (NDTV 2020), a start-up that enables closed-loop recycling of rigid plastics to match the quality of virgin plastic. At about the same time, Hindustan Unilever announced that they had made 100 million bottles (Mint 2021) from recycled plastic granules supplied by Banyan Nation for use by several of their brands. Most recently, in September 2021, HUL also unveiled a bottle made of 50% recycled plastic for one of their brands (Mint 2021).

These percentages of recycled content in packaging and the volumes delivered are major milestones in packaging recycling in India. They represent a design, technical, and value-chain partnership accomplishment that requires tremendous synergy between all actors. For enabling such usage of recycled content in plastic packaging, the entity supplying the recycled resin has to create a resilient and reliable supply of raw materials (plastic waste) in order to consistently produce recycled resin in the quantities and standards that are demanded by the large enterprises. This has historically been challenging to achieve with the fragmented supply chain that exists for secondary raw materials in the country and the dependence on the informal sector. Start-ups like Banyan Nation have successfully demonstrated the use of digital platforms to achieve previously challenging levels of supply-chain integration, which have been critical to their success and the credibility they have built with large-scale partners. Platforms like these connect thousands of last-mile collectors (frontline waste pickers) from the informal sector. For Banyan Nation, their digital platform was also critical in enabling them to partially limit the shock of the pandemic disruptions. It aided in the quick resumption of raw material procurement from their supply network as lockdowns began to ease across India. (Chapter 6 discusses Banyan Nation and its business model in a detailed case study.)

8.2.2 Financing Innovation in Plastic Recycling

While concerns escalated on the state of the plastic recycling industry after COVID-19's impact, important advances were made in the space of financing innovation in plastics recycling. In December 2020, Circulate Capital, an investment management firm specialising in advancing circular economy solutions, announced investments to the tune of USD 39 million (under the Circulate Capital Ocean Fund) in six companies in India (Press Trust of India 2020). This investment portfolio reflects a much-needed value-chain approach for investing in the Indian plastic recycling ecosystem, covering the scaling of collection and sorting operations, the digitisation of the waste-management ecosystem, and several kinds of upcycling technologies. Mirova Natural

Capital also invested USD 2 million (Pathak 2020) in Bangalore-based social enterprise Plastics for Change (through The Althelia Sustainable Ocean Fund) to scale up their work in India and across markets in South-East Asia.

The emergence of specialist investment funds that can provide risk capital to Indian entrepreneurial ventures in the plastic recycling sector, as well as the growth of these Indian ventures over the past few years to a stage where they can attract such capital, are important developments in the sector. They mark a turning point in what has long remained a sector that struggles to prove itself as investment-worthy.

8.2.3 Industry Partnerships and Commitments on Plastic Recycling

With rising consumer awareness around issues of sustainability and waste, several domestic and global consumer brands in India also made commitments to using recycled plastic to replace virgin plastic. The launch of the India Plastics Pact, the first of its kind in Asia, is expected to augment the recycling ecosystem in the country. The Plastics Pact was launched by World-Wide Fund for Nature-India (WWF India) and the Confederation of Indian Industry (CII) and supported by the Waste and Resources Action Programme (WRAP) and UK Research and Innovation (UKRI) (India Plastics Pact 2021). The Pact brings together businesses, the government, nongovernmental organisations (NGOs), nonprofits, and other stakeholders to achieve certain time-bound targets around

o making plastic packaging reusable or recyclable;
o effectively recycling plastic packaging;
o incorporating recycled content across all plastic packaging.

8.2.4 Policy Impetus to Plastic Recycling

The period since the pandemic began has also seen some movement from State and Central Governments that can boost the struggling plastic recycling industry. For instance, in November 2020, the Maharashtra State Government directed (through conditions in 'Consent to Operate') that plastic packaging for non-food/non-pharma applications by producers and leading brand owners should contain a minimum of 20% recycled content (MPCB 2020). The Government has also begun looking into how recycled content can be promoted in non-packaging plastic applications, including manufacturing of items such as moulded items, mats, buckets, agricultural pipes, and fittings. The latest notification on Extended Producer Responsibility issued by the Ministry of Environment, Forest and Climate Change (MoEFCC), Government of India, also includes targets for recycling and for the use of recycled content in plastic packaging (MoEFCC 2022), based on the category of plastic packaging. Such initiatives can help the plastic recycling industry by inducing demand and stabilising the market for recycled plastic resins.

The Indian plastic recycling industry is still heavily reliant on downcycling and has a long way to go to achieve closed-loop recycling. Systemic changes along the entire plastic production and consumption phases are required to enable this transition – these include measures to limit the sheer variety of plastics in use, the standardisation and harmonisation of plastic composition for various use cases, designing out barriers to recyclability, and improving the quality of the plastic waste available for recycling. The devastation to the industry needs to be looked at as a potential opportunity to rebuild the plastic recycling sector while incorporating greater circularity.

The positive developments described in the sections above provide some hope and cues for rebuilding the plastic recycling sector. While these fledgling developments currently influence only a small percentage of the plastics recycling value chain in India, there is a strong need to capitalise on these movements, and draw lessons from the implementation of these newer business models, policies as well as financing and collaboration initiatives, that have bolstered the industry despite the challenges of the pandemic. The pandemic has also highlighted the need to thoroughly map the plastic recycling ecosystem in the country, gather credible and actionable data, and make this data available to all stakeholders. These measures will become even more relevant as India moves towards phasing out SUPs by July 2022. In transitioning the plastic recycling ecosystem, consideration will have to be given to how the economics of plastic recycling will work when wages and health conditions in the sector are enhanced, and as technology is upgraded for improving recycling outcomes.

8.3 DISTRIBUTED MANUFACTURING AND LOCALISED PRODUCTION

The circular economy framework implores us to fundamentally rethink how we make and consume goods in order to maximise efficiency of resource use, minimise waste, and use resources regeneratively. In addition to strategies such as resource recovery and recycling, remanufacturing, providing access over ownership, and employing circular design, a circular economy also encourages distributed manufacturing and localised production. In distributed manufacturing, production takes place at a more localised scale and is closer to the final consumer. This is in contrast to current manufacturing models, which are reliant on high degrees of centralisation, where raw materials are transported to a central plant for production, and finished goods from the central plant are distributed across vast distances to reach the final consumer. Overall, local production and small-scale, distributed manufacturing is expected to lead to optimisation of resources as well as energy for transportation. In the long term, it can also create local circulation of nutrients and selective extraction of suitable local resources from the natural ecosystem.

Our current global economy is only 8.6% circular (Haigh et al. 2021). In a post-pandemic world that aims for greater circularity to meet climate change goals, decentralisation of the global production network and shortening of supply chains through distributed manufacturing and localised production is a critical strategy to consider. A circular economy will have to find a balance between production and closing of loops at a global level (for example, for products with highly complex supply chains) versus at a more regional or local level. Consideration will also have to be given to issues that will arise in rebalancing global trade and challenges around a just transition.

The pandemic-associated conditions threw into disarray highly globalised supply chains which have been increasingly optimised for efficiency and cost over the last few decades, most often at the expense of resilience. According to a new report from the McKinsey Global Institute, supply-chain disruptions that last a month or longer are projected to occur roughly every 3.7 years (Lund et al. 2020). In this context, it becomes increasingly imperative for businesses to build resilience into their procurement and production, including through decentralised models.

The M19 community-based initiative by Maker's Asylum (India's first community maker space) serves as a great illustration of the potential of decentralisation – specifically, how the distributed manufacturing model helped withstand supply-chain uncertainties and movement restrictions in the early days of the nationwide lockdown. M19's distributed manufacturing model relied on decentralised procurement of materials, production of face shields, and their distribution through localised hubs. In 49 days, the initiative made over 1 million face shields through the mobilisation of a maker community spread across 42 cities/towns/hubs in India (Maker's Asylum 2021), a feat that was not possible even for the bigger and more established manufacturers. The distributed manufacturing model also allowed Maker's Asylum to run through rapid iterations and prototyping to come up with an eco-friendly design. The distributed model also ensured that Maker's Asylum chose material for the shields that was ubiquitously available across India, inexpensive to procure, fit for purpose, and easily scalable in the context of high demand for PPE and broken supply chains and distribution infrastructure across the country.

Decentralisation and localisation of production centres could also play a key role in enabling models that are vital to *closing the loop* in a circular economy, including reuse, repair, refurbishing, and remanufacturing services; resource recovery; and mining for secondary raw materials. Since these models are highly reliant on *reverse* networks and logistics, they are potentially easier to establish and maintain at local or regional scales, and are likely to be more economically viable at least for certain kinds of products and use systems (for instance, glass bottles in local reuse systems). They also have the potential to create more jobs at the local level, aided by relevant upskilling of the workforce.

In addition to the supply-chain vulnerabilities exposed due to COVID-19, other events and trends also point to the potential for a high degree of decentralisation being on the horizon. The Ever Green container ship's blocking of the Suez Canal led to global trade losses to the tune of US\$9.6 billion per day (Baker et al. 2021), highlighting the limitations of the centralised nature of our current supply chains. Manufacturing models such as delayed differentiation (where customisation of products takes place closer to the consumer) are projected to become mainstream in the near future. Innovative models are also emerging for closing loops locally, such as micro-factories that can enable decentralised solutions for the recycling of e-waste (Mehra 2020). It is vital for businesses and governments to understand how the trends in distributed/decentralised manufacturing can be aligned with the emergence of new models for closing loops, and be leveraged towards a systemic circular transformation.

8.4 CONCLUSION

The conditions imposed by the pandemic have served as an acid test for the manner in which our predominantly linear global economy functions. In India, these conditions have had an overall negative impact on the plastics recycling industry, with many recyclers shutting down or operating at low capacities. However, the same period also witnessed some positive developments for this sector (particularly plastics packaging), including proving the resilience of some closed-loop recycling models, attracting financing for start-ups, strengthened commitments from FMCG brands, and policy impetus to plastic recycling. As India moves towards a phase-out of SUPs in July 2022, there is a need to build a well-functioning plastic recycling sector engaged in high-value recycling that can deal with the recycling of potential alternatives to SUPs.

The pandemic also positively influenced the uptake of subscription models for mobility, accelerating changes in consumer behaviour that would have otherwise taken much longer to materialise. This period also brought to light the potential of decentralised production as a model for resilience, as demonstrated by the M19 initiative.

These developments served to highlight several aspects that are critical to the CE transition in India, which may also offer lessons for the global South:

- *Innovation in business models* by start-ups as well as community initiatives like M19 have played a major role in accelerating the move towards a circular economy, even through the chaos of the pandemic. The fact that these organisations were able to innovate and attract funding and partnerships in the midst of COVID-19 is a positive signal. It is potentially indicative of their ability to survive the bigger challenges and rapidly shifting landscape that lie ahead in the CE transition. Mechanisms to further encourage the development and scaling of such models is a priority for the CE transition.
- *Digital technologies* have also proven to be critical in supporting these new business models. Digital technologies underpin models like subscription, and they provide ease of access for users while allowing for better operational management and greater reach for the businesses providing these services. Digital platforms have also proven to be instrumental in integrating informal supply networks to create resilient supply chains for raw material for plastic recycling. Decentralised production models are also heavily reliant on digital technology for information transfer and coordination amongst the distributed production networks. Leveraging the immense digital innovation capabilities in India's workforce is vital to ensure the continued development of customised solutions that can address challenges and opportunities in the country's CE transition.
- *Livelihoods:* Rebuilding the Indian plastics recycling industry towards closed-loop recycling models has tremendous potential to improve livelihoods of the informal-sector workers that form its backbone. Decentralised models of production along with local circulation of products and materials through repair, reuse, and refurbishment have the potential to create high-skilled jobs regionally.
- *Systemic planning and transitional strategies*: The system-wide shock resulting from the COVID-19 pandemic has resulted in a number of CE innovations and the accelerated uptake of some existing innovations. While innovations are critical to the CE transition, there is a need to direct efforts towards gathering credible

data and holistically assessing their impact and consequences on the entire ecosystem. For instance, subscription models can lead to increased number of vehicles for personal mobility and shift users away from public transport, and models like decentralised production can entirely shift the trade balance between countries. Proper systemic planning is required to ensure that innovations do not lead to unintended consequences and burden-shifting to different parts of the lifecycle or to different regions of the world. As highlighted in the case of plastics recycling, systemic planning is also vital to ensure that upstream interventions towards circularity complement and improve overall impact of downstream innovations. The circular transition is also likely to have fallouts such as job redundancies in some sectors/geographies, a loss of business opportunities due to higher sustainability standards of products, etc. Systemic planning should also take into account the transitional strategies needed to deal with some of these fallouts and to ensure a just transition.

REFERENCES

Accenture. 2018. "Accelerating India's circular economy shift: A half-trillion USD opportunity", FICCI CES, 2018. https://ficcices.in/pdf/FICCI-Accenture_ Circular%20Economy%20Report_OptVer.pdf.

Accenture. 2020. "Strategies for sustainable plastic packaging in India: A USD 100 billion opportunity till 2030", FICCI CES, 2020. https://ficci.in/spdocument /23348/FICCI-Accenture-Circular-Economy-Report1.pdf.

Aravind, Indulekha. Economic Times. 2020. "Another pandemic: India's fight against single-use plastic falls victim to Covid", Economic Times, October 24, 2020. https://economictimes.indiatimes.com/news/politics-and-nation/another-pan-demic-indias-fight-against-single-use-plastic-falls-victim-to-covid/articleshow /78848847.cms.

Baker, James, Eric Watkins, David Osler. Lloyd's List. 2021. "Suez Canal remains blocked despite efforts to refloat grounded ever given", Lloyd's List, March 24, 2021. https://lloydslist.maritimeintelligence.informa.com/LL1136229/Suez -Canal-remains-blocked-despite-efforts-to-refloat-grounded-Ever-Given.

Bond, Kingsmill, Harry Benham, Ed Vaughan, Lily Chau. Carbon Tracker. 2020. "The Future's not in plastics", Carbon Tracker, September, 2020. https://carbon-tracker.org/reports/the-futures-not-in-plastics/.

Brock, Joe. 2020. "The plastic pandemic: COVID-19 trashed the recycling dream", REUTERS, October 5, 2020. https://www.reuters.com/investigates/special -report/health-coronavirus-plastic-recycling/.

Choubey, Sonika. YouGov. 2021. "Most urban Indians wanting to buy a car have delayed their plans by four months or more", YouGov, May 11, 2021. https://in.yougov.com/ en-hi/news/2021/05/11/most-urban-indians-wanting-buy-car-have-delayed-th/.

Deloitte. 2021. "2021 global automotive case study - Global focus countries", Deloitte, 2021. https://www2.deloitte.com/global/en/pages/consumer-business/articles/ global-automotive-trends-millennials-consumer-study.html.

Dharmadhikari, Sanyukta. 2020. "The pandemic has dented the war against plastic in Bengaluru", The News Minute, September 17, 2020. https://www.thenewsminute .com/article/pandemic-has-dented-war-against-plastic-bengaluru-133296.

Ellen MacArthur Foundation. 2016. "Circular Economy in India: Rethinking growth for long-term prosperity", Ellen MacArthur Foundation, December, 2016. http:// www.ellenmacarthurfoundation.org/publications.

Ellen MacArthur Foundation. 2021. "A solution to build back better: The circular economy" Accessed November 20, 2021. https://emf.thirdlight.com/link/ w68inpdhttkd-71ev4i/@/preview/1?o.

GA Circular. 2020. "Safeguarding the plastic recycling value chain: Insights from COVID-19 impact in South and Southeast Asia", GA Circular, August 2020. https://1b495b75-5735-42b1-9df1-035d91de0b66.filesusr.com/ugd/77554d_646 4ccce8ff443b1af07ef85f37caef5.pdf.

Haigh, Laxmi, Marc de Wit, Caspar von Daniels, Alex Colloricchio, Jelmer Hoogzad. Circle Economy/Shifting Paradigms. 2021. "The Circularity Gap Report 2021", Circle Economy, January, 2021. https://assets.website-files.com/5d26d80 e8836af2d12ed1269/60210bc3227314e1d952c6da_20210122%20-%20CGR %20Global%202021%20-%20Report%20-%20210x297mm.pdf.

India Plastics Pact. 2021. "How will it work?", Accessed on December 12, 2021. https://www.indiaplasticspact.org/.

KPMG India. 2020. "The future of mobility in post – COVID-19 India", KPMG India, September 2020. https://assets.kpmg/content/dam/kpmg/in/pdf/2020/09/the -future-of-mobility-in-post-covid-19-india.pdf.

Lidhoo, Prerna. Fortune India. 2020. "COVID-19 has been a boon for car rentals: Zoomcar", Fortune India, December 12, 2020. https://www.fortuneindia.com/ venture/covid-19-has-been-a-boon-for-car-rentals-zoomcar/104938.

Lund, Susan, James Manyika, Jonathan Woetzel, Edward Barriball, Mekala Krishnan, Knut Alicke, Michael Birshan et al. McKinsey Global Institute. 2020. "Risk, resilience and rebalancing in global value chains", McKinsey Global Institute, August, 2020. https://www.mckinsey.com/business-functions/operations/our -insights/risk-resilience-and-rebalancing-in-global-value-chains.

Maker's Asylum. 2021. "M-19 Shields for COVID-19", Accessed on December 12, 2021. https://makersasylum.com/m19-shields/.

Mehra, Preeti, Hindu Business Online. 2020. "Microfactories will build a circular economy for waste management: Veena Sahajwalla", Hindu Business Online, February 26, 2020. https://www.thehindubusinessline.com/specials/clean-tech /microfactories-will-build-a-circular-economy-for-waste-management-veena -sahajwalla/article30915503.ece.

MG Motor India. 2020. "MG Motor India partners with Zoomcar for vehicle sub- scriptions", MG Motor India, August 18, 2020. https://www.mgmotor.co.in/ media-center/newsroom/mg-motor-india-partners-with-zoomcar-for-vehicle -subscriptions.

Mint. 2021. "HUL moves to recyclable bottles for Surf Excel Matic Liquid", Mint, September 14, 2021. https://www.livemint.com/companies/news/hul-moves -to-recyclable-bottles-for-surf-excel-matic-liquid-11631618830756.html#box _11631618830756.

MoEFCC. 2022. "Guidelines on Extended Producer Responsibility for plastic packaging: Notification", Ministry of Environment, Forest and Climate Change, February 16, 2022. https://egazette.nic.in/WriteReadData/2022/233568.pdf.

Mohile, Shally Seth. Business Standard. 2021. "India auto sales down 13%, industry retreats by six years in pandemic", *Business Standard*, April 12, 2021. https:// www.business-standard.com/article/companies/indian-auto-sales-down-13 -industry-retreats-by-six-years-in-pandemic-121041200408_1.html.

MPCB. 2020. "Sub: Amendment in consent to operate of leading brand owners and plastic producers for use of recycled plastic in plastic packaging and plastic items", Maharashtra Pollution Control Board, November 18, 2020. https://www .mpcb.gov.in/sites/default/files/standing_orders/CircularAmendmentinCtoO leadingBrandOwners02032021.pdf.

NDTV. 2020. "Dettol unveils one-of-a-kind initiative of handwash bottles made from 100 per cent recycled material", SwacchIndia NDTV, October 2, 2020. https:// swachhindia.ndtv.com/video-details-page/dettol-unveils-one-of-a-kind-initiative-of -handwash-bottles-made-from-100-per-cent-recycled-material-562185/.

Pathak, Kalpana, Mint. 2020. "Plastics For change raises $2 million from Mirova Natural Capital", Mint, May 30, 2020. https://www.livemint.com/news /india/plastics-for-change-raises-2-million-from-mirova-natural-capital -11590820105222.html.

Pathania, Ramesh. Mint. 2021. "Most customers are unaware of vehicle subscription model, finds Car Dekho survey", Mint, April 16, 2021. https://www.livemint .com/news/india/most-customers-are-unaware-of-vehicle-subscription-model -finds-car-dekho-survey-11618558998730.html.

Peng, Yiming, Peipei Wu, Amina T. Schartup, Yanxu Zhang. 2021. "Plastic waste release caused by COVID-19 and its fate in the global ocean", *Proceedings of the National Academy of Sciences of the United States of America*, Vol. 118 No. 47, e2111530118, October 6, 2021. https://www.pnas.org/content/118/47/e2111530118.

PlastIndia Foundation. 2019. "Indian Plastics Industry Report 2019", PlastIndia Foundation, 2019. https://www.plastindia.org/plastic-industry-status-report.php.

Press Trust of India. 2020. "Circulate Capital commits $19 million investment in 4 waste management cos in India", *Financial Express*, December 10, 2020. https:// www.financialexpress.com/industry/circulate-capital-commits-19-million -investment-in-4-waste-management-cos-in-india/2147348/.

Puthran, Nikhil, CarWale. 2021. "CarWale 2021 IACC Survey: 60 per cent respondents to consider online purchase", *CarWale*, February 25, 2021. https://www .carwale.com/news/carwale-consumer-survey-indian-automotive-consumer -canvas-iacc-2021/.

Raveendran, Govindan and Joann Vanek, WIEGO. 2020. "Informal workers in India:A statistical profile", Women in Informal Employment: Globalizing and Organizing, August, 2020. https://www.wiego.org/sites/default/files/publications /file/WIEGO_Statistical_Brief_N24_India.pdf.

Revv. 2021. "Mahindra subscription", Accessed December 9, 2021. https://www.revv .co.in/mahindra-subscription.

Singh, Abhilasha, ExpressDrives. 2021. "Car subscriptions to become more popular than car buying: Sakshi Vij, Myles", *ExpressDrives*, September 14, 2021. https:// www.financialexpress.com/auto/car-news/car-subscription-rental-sakshi-vij -myles-cars-zoomcar-avis-india-maruti-hyundai-tata-mahindra/2326201/.

Suratman, Nurluqman. ICIS. 2020. "SE Asia, India plastic recycling weighed down by pandemic, low oil prices", *Independent Commodity Intelligence Services*, August 20, 2020. https://www.icis.com/explore/resources/news/2020/08/20 /10543073/se-asia-india-plastic-recycling-weighed-down-by-pandemic-low-oil -prices/.

Tata Motors. 2021. "EV Subscription", Accessed December 9, 2021. https://evsubscrip-tion.tatamotors.com/.

Transport & Environment. 2019. "Europe's giant 'taxi' company: Is Uber part of the problem or the solution?", European Federation for Transport and Environment AISBL, November, 2019. https://www.transportenvironment.org/wp-content /uploads/2021/07/T&E_Europe%20s%20giant%20taxi%20company%20is %20Uber%20part%20of%20the%20problem%20or%20the%20solut...%20(1)_1 .pdf.

Vasanth, Pon B.A. 2021. "Single-use plastics back in use", *The Hindu*, July 31, 2021. https://www.thehindu.com/news/cities/chennai/single-use-plastics-back-in-use/ article35642423.ece.

Venkata-Subramani, Jesse Roman. 2020. "The coronavirus response in India – World's largest lockdown", *The American Journal of The Medical Sciences*, Volume 360, Number 6, 743, December, 2020. https://www.ncbi.nlm.nih.gov/pmc/articles/ PMC7405894/.

Zoomcar. 2021. "Zoomcar mobility services", Accessed December 9, 2021. https:// www.zoomcar.com/zoomcar-mobility-services.

Index

For Product Safety Concerns and Information please contact our EU
representative GPSR@taylorandfrancis.com
Taylor & Francis Verlag GmbH, Kaufingerstraße 24, 80331 München, Germany